The Camerons of Glenspean

The Camerons of Glenspean

The family behind Meredith Dairy:
Five generations of Australian initiative and innovation

Neil Gordon Cameron

Contents

PART ONE	7
Introduction	9
1 The property Glenspean	13
A sidelight	14
2 Neil Wilson Cameron	15
Stockman, soldier and family man	15
Kay Cameron's kitchen, 1925	21
Refrigeration	23
Electricity supply	24
Communication	24
Water supply and sewerage	25
3 Neil's and Kay's antecedents	30
Alexander Clayhills Cameron	30
The spread of the Cameron vine	33
Kay Browne's wedding veil, whereon there hangs a tale	34
4 Neil Wilson Cameron's innovative thinking	37
Shelter belts	38
Cattle	39
Sheep	39
Finance	40
PART TWO	43
5 Neil Gordon Cameron: early days, growing to manhood after a shaky start	44
Woodburn Creek State School	44
Park Street	45
Boarding school	46
Water and windmills and rowing	48
6 University life	56

7	**Agricultural scientist Neil: returning to the practicalities of farming**	**59**
	Cattle and haymaking: the Roto-Baler	59
	A peep backwards in time	60
	Hay for the big-heads as John Cameron called them	*61*
	First pole for new hay shed, 1950	*62*
	Stones	65
	Sheepman Neil: early mustering and later	65
	Rabbiting	67
	Map: Glenspean 1924	*68*
	Map: Glenspean 1975	*69*
8	**Using my agricultural degree**	**72**
	Cattleman Neil and Prof Terry Robinson	*72*
	Terry Robinson's world-famous Glenspean Experiment 1953–58	*76*
	Breeding value	*77*
	Nutrition	*77*
	Other factors	*77*
	Growing cattle store energy	80
	Results	*81*
9	**Fireman Neil**	**84**
	Black Wednesday, 22 February 1967	84
	Aftermath	90
	Thursday 23 February 1967	90
	Friday 24 February	91
	Saturday 25 February	91
	Sunday 26 February	92
	Monday 27 February	92
	Friday 3 March	92
	Lethbridge Fire, Wednesday 8 January 1969	93
	Lal Lal Grassfire, 22 January 1967	*95*
10	**Afterthoughts**	**96**
11	**Pioneer Park**	**98**
	Epilogue	**108**

Part One

Neil Wilson Cameron:
The central figure of our story, who bought *Glenspean*

Sandy and Julie Cameron

Introduction

In 2018 the Camerons of *Glenspean* would most properly be associated with Julie and Sandy Cameron who are the strength behind the ever-growing Meredith Dairy. It is based at a *Glenspean* much enlarged from the property that my father bought; yet it would not exist but for the work of those who have gone before, firstly Neil and Kay, then June and me – Neil Jr. Meredith Dairy was fledged from the Glenspean Partners nest, but the fledgling is now a soaring eagle providing financial viability to me, Neil, the part-owner of the land. All in all, Meredith Dairy is a great intergenerational family edifice built on solid foundations.

Sandy and Julie's business encompasses four sheep- and goat-milking farms with the on-farm cheese and yoghurt factory also supplied by several outside milk suppliers under contract. The main farm supplements the factory, supplying grazing, grain, silage and straw. The whole is backed up by an extensive marketing network. Sandy's and Julie's cheeses and yoghurts augment larders around half the world. Theirs is a story still to be told as it unfolds. It is one for their family to tell in the future. I am only able to relate some of what has gone before.

Glenspean has a long history of debt not helped by drought and the failure of the Reserve Price Scheme for Wool. At first, money was tight due to shortage of capital, and the partnership was floundering due to the time taken for Sandy to bring its management up to speed and for the Meredith Dairy to get up steam.

Julie and Sandy have had their setbacks. Father-in-law Neil has not always been helpful, as when trying to carry two trays of jars to the health food shop in Daylesford I dropped the week's produce onto the pavement; or when unloading a pallet of jars I pressed the control on the front-end loader the wrong way and sent the jars crashing to the ground. A real smash hit! I am sorry Julie. You have always been so kind, forgiving and welcoming towards me, ready to give or receive a kiss.

To balance the ledger I mixed most of the concrete for the first milking platform and subsequently built the elevated roofed batten space for the sheep waiting to be milked. I used all second-hand materials; money was so tight. I had four large ironbark poles with which I had intended to make a sheep bridge over the all too frequently swollen creek. With a chainsaw I halved them longitudinally to make the eight uprights supporting the roofed batten space, the entry to the initial sheep milking platform, now demolished. When completed I concreted the rock entry ramp. As partner in the parent Glenspean Partners, June did many an errand in Ballarat, whilst in addition to the normal farm work, I did weekly trips hither and thither, distributing anything that should have brought in some cash.

June being no longer active, and me in various stages of decrepitude, we oldies are about to get out of the picture completely, except for helping with several sheep mustering jobs most weeks, and being determined, if permitted, to swing a gate on my ninetieth birthday later this year, 2018. Our retirement from the partnership will leave Julie, Sandy and family free to build the business to new heights, as the spirit moves them.

The dairy provides welcome employment opportunities. Locally, the many who are willing to put their shoulders to the wheel find themselves treated as colleagues by Sandy and Julie with their hands-on approach. Globally, those offering their expertise come rushing from every continent, except Antarctica, to keep the ball rolling.

In 2018, when thinking of the Camerons of *Glenspean*, the focus tends to be on our son Sandy and his wife Julie, but Sandy's two brothers, Simon and Hamish, have done equally well, pulling themselves up by the bootstraps.

Simon has recently taken on the task of converting a thousand hectares of bare degraded pasture into a productive property, grazing a superfine Merino flock selected over 30 years for low fibre diameter and high fleece weight. Wendy named the property Black Swan Crossing for the large pools of water in the ephemeral Mustons Creek, which snakes through the property. It is south-west of the one-horse hamlet of Hexham and, more recognisably for most, distant 15 kilometres west of Mortlake in Victoria's Western District. Hexham is a 'watering hole' on the Hopkins River: it has a pub!

Simon's first job on the bare land was to build a set of sheep yards, allowing him to bring his fine wool Merino flock from *Loma*, his former property of less than half the size in the north-east. The yards were followed by a modern shearing shed with 'raised board'. The yards have since been augmented with sheep handling equipment, enabling Simon to do his own drenching, crutching and jetting.

Then he built his four-bedroom house with only a minimum of help from electricians and plumbers as required by the statutory authorities. Environmentally friendly, it faces north, its length aligned east-west. It is heavily insulated and clad with corrugated iron.

Whilst doing all this Simon has taken the first steps towards improving the pastures by fertilising them with superphosphate. In addition he has taken time to plant shelter belts and with Wendy to begin a garden. With a science degree from Melbourne University, Wendy was Brown Bros Chief of Winemaking before retirement. She still acts as consultant to the Brown Bros winery based at Milawa in Victoria's north-east, and in response to the warming climate was responsible for their decision to buy a vineyard in cooler Tasmania. She is now pursuing a doctorate on the effects of climate change on winemaking.

Simon has become so fluent in German that he listens to the news and reads books in German. He has taken his family on several trips to Germany.

Beginning with a modest loan, now repaid, Hamish has likewise shown enterprise by more than doubling the physical size and scope of his West Queanbeyan Veterinary Hospital. Still on the up and up, he has been voted the third most popular vet in Australia. Simon is a Master of Veterinary Science. Sandy has added a doctorate to his veterinary degree. All three were honours graduates.

Our daughters are concerned for the welfare of people. Jane, with three children of her own and now married to John Young, is a physiotherapist specialising in women's health. She has acquired a postgraduate certificate in pelvic floor rehabilitation and incontinence. Kate has a master's degree gained through her research into the effects of stress on the immune system. She is a sports physiologist, and is married to Michael Greenwood.

I am extremely grateful for Mum's and Dad's decision to buy *Glenspean*, and a quarter of a century later to bring my brother John and me into partnership with them, eventuating in our ownership of their property. Ownership would not have happened for me but for the generosity of my mother and John in rearranging titles: returning from university I found myself managing my end of the farm but owning nothing of it. John's continuing generosity of spirit enabled us to work together as joint managers for 32 years until we dissolved the partnership in 1984 to allow our children a leg-in. As the ones now in the money, thanks to their dairy, Sandy and Julie have shown me a similar generosity. Built up by five generations, from my great-grandparents' time, all showing initiative and perseverance and now going into the sixth generation, we are the Camerons of *Glenspean*.

The Meredith Dairy enterprise would not exist but for the vision of Sandy's grandfather, Neil Wilson Cameron, who bought the home property, *Glenspean*, 94 years ago. It is with him that we take up *our* story. According to my mother, Dad, with his fiancé Kay, sat on a rock overlooking the creek that cuts through the place and made the decision to buy *Glenspean*, because it was within a day's drive of *Kirkbank* in Malvern, Melbourne's eastern suburb, the home of Kay's father, who was feeling his age.

As our story of Dad proceeds, *Glenspean* was then covered in stones and tussocks, but according to Mum, 'Dad thought he could make a go of it'. Starting with very little capital he most certainly had to work hard to succeed, as his grandson Sandy with Julie and Sandy's siblings have all had to do.

June and Neil Cameron, celebrating 60 years of marriage.

1 The property Glenspean

Glenspean was named by George McNaughton, whose mother had been a Cameron.

In 1852 Duncan McNaughton and his wife Margaret came to Australia aboard the ship *Clifton* with their children, including George and Donald. They stayed in Melbourne for three years then settled at Inverleigh in 1856 and set up a blacksmith shop. Their sons, George on *Glenspean* and his brother Donald on *Fernhill,* had taken up some of the 80-acre blocks thrown open for selection when 'Poor Man' Read's grazing licence was cancelled as the gold rush petered out. 'Gentleman' Reid, as he was known in Meredith, retained his adjoining grazing licence, Woodbourne.

George built his four-roomed weatherboard house, *Glenspean*, and with it a bluestone dairy. He became well known for his cheese. He was also a long-serving member of the original Meredith Shire Council. He married Jesse McCallum in 1875. They had eight children. (Marg Cooper's *Woodburn Revisited* 2007 citing *Founders of Australia*.)

George Frederick Read, the son of a Tasmanian farmer, in 1837 at the age of 17 landed his sheep at Point Henry or thereabouts and drove them northwards until eventually squatting beside the Cargerie Creek, a dozen kilometres west of the future Meredith township. Read was armed with a grazing licence, unlike my maternal great-grandfather, John Brown, who tried to squat on the Bellarine Peninsula without one. Read later obtained a pre-emptive right for the square mile (640 acres; 259 hectares) surrounding his homestead. His death coinciding with the clamour for land following the decline in the Ballarat gold boom, GF Read's grazing licence was cancelled, which was when George McNaughton became the owner of the Crown allotments on which the *Glenspean* homestead now stands.

A sidelight

Clarrie Meredith, a doyen of the Sorrento Presbyterian Church, related to me stories of how Meredith had got its name, stories which he had heard on his grandfather's knee. The brothers Meredith had landed at Adelaide, proceeded to Geelong, and upon the discovery of gold at Buninyong and Ballarat, walked in that direction. The younger brother Charles apparently became sick of walking and opened an inn where the track crossed the Coolebarghurk Creek. That resting point for weary travellers became known as Meredith's and subsequently the township as 'Meredith'. Officially, Meredith township was named from a von Steiglitz connection in Tasmania

Clarrie told me another anecdote: on the voyage to Australia the brothers decided to cut a dash and went on deck wearing their top hats, which immediately were blown into the sea. I lost my top hat into Lake Wendouree when I decided that I should be properly dressed for the opening of the Ballarat Yacht Club sailing season. I had previously joked with other club members that racing on the lake was so tame compared to Port Phillip Bay that one could do so in one's dinner suit without it getting wet; so for the club opening in 2013 I donned the morning coat and striped trousers which I had worn at my brother's marriage in February 1952. To my delight Deputy Mayor Samantha agreed to take part with me in the sail-past which concluded the opening ceremony. My grey top hat was knocked off by the boom when we went about, to be rescued by the safety boat – to much mirth.

When George McNaughton's *Glenspean* was eventually sold, the new owner moved the house 150 metres eastwards, to a less stony site, its four rooms forming the centre section of *Glenspean* as I remember it in my youth. According to Keith McColl, the enlarged house was built in 1918, for JB Were, I think he said. There have since been several additions and alterations, each attempting to cope with the unfortunate north-south axis on which the enlarged house was built. There had been the common but mistaken belief that Meredith had a hot climate: I suppose that, compared to British winters, that notion was true. Consequently the rebuilt *Glenspean* had a Queen Anne front to the south, graced with a wide verandah and stained glass in the windows. It had a maryanne kitchen in the north, without a verandah: it mattered not if the mary got hot (*mary: colloq. a woman, i.e. the cook or kitchen maid*). It was built as a wedding present for JB Were's daughter. The marriage never took place and the ownership of *Glenspean* then passed through several hands, including the Bells of Bells Socks, before being bought by Mum and Dad.

2 Neil Wilson Cameron

Stockman, soldier and family man

Dad may well have also been influenced to buy the property *Glenspean* by a romantic attachment to the name 'Glenspean' as Cameron country in Scotland. Strictly speaking Cameron country was not the 'Glen' of 'Glenspean' but the gloaming or flat country beside the Spean River, with the Cameron lands extending north across the Great Glen to Lochaber. The clan had no legal title but the clansmen held the land by right of historical occupation. That was before Bonnie Prince Charlie's foolish tilt at the English throne, which led the clans into disaster, culminating in the battle of Culloden. After the clansmen's claymores had been confiscated by the Duke of Cumberland, the 'Butcher', they were evicted by the clan chief who no longer had need of clansmen without their claymores. He went to the Royals in London, Queen Mary II and William, and claimed a Crown title to the land which had formerly been jointly owned by the clan as a whole.

My dad was the third of six brothers who, with their parents, lived at *Balgartnie* in the heights of Heidelberg, an outer suburb of Melbourne. The boys were Alistair, Colin, Neil, Evan, Norman and Malcolm. They were a close family, and remain so into the fourth generation, despite geographical separation from Hay in New South Wales to Portland on the Victorian west coast and Naracoorte in the south-east of South Australia. They had lost their much loved brother Colin at Gallipoli; but the five survivors all worked up from almost nothing to have their own substantial grazing properties.

From Heidelberg the boys rode their ponies across the Yarra River to Scotch College in Kew. I am told that the College gave a ten per cent discount on fees for a boy having a brother at the College, with the result that the youngest received his education half-price. Their father Wilson was a stock agent at the central Victorian selling centre, Newmarket, close by Flemington where the 2017 Melbourne Cup had just been run, as I first wrote this. They spent much of their holiday time droving sheep and cattle to and fro at the saleyards, from which Dad received his initial training in stockmanship. Dad left school at the end of the Intermediate year, year ten, having as he said learned but one useful thing, 'Agricola – a farmer'. The boys paired off to run small dairy farms around Beveridge, north of Melbourne. They and their families maintain that commitment, with the most recent triennial reunion in April 2018.

With the regimental number of 300, Neil's next older brother Colin was one of the first to join up on the outbreak of the Kaiser's war. He joined the Eighth Light Horse, but upon arrival in the Middle East the Light Horsemen were de-mounted and sent to Gallipoli where Colin was killed in the futile feint on the Nek. Recently promoted Sergeant Major, he was responsible for mustering troops for the first wave of the attack in which he himself was killed.

In Uncle Malcolm's old age I asked him why Colin's name was never mentioned in the family. 'It was too painful,' replied Malcolm, who also made the comment that Colin had been a fine horseman.

Left: Neil Wilson Cameron, appalled at Hitler's ravages, rejoined the army in 1940

Cameron family c.1914 apparently at Highpark, Lancefield; back row standing L–R Colin and Alasdair; centre row seated L–R Evan, Charlotte, Wilson, Neil; front row on ground L–R Malcolm and Norman

Upon Colin's death Dad joined up. Evan had already done so and, following the ninth of July disaster resulting in Colin's death, he was sent to Gallipoli as a reinforcement. Dad joined the artillery with which he spent three years in France, 1916–18, where he was joined by Alistair and Evan.

A signal-man, Dad was responsible for keeping open lines of communication to the guns from the O-Pip, the observation post where the fall of the shells was observed. The battle reached a critical stage in 31 August – 1 September 1918. During a particularly heavy German bombardment Dad kept open those lines of communication, for which 'bravery and devotion to duty' they conferred on him the Military Medal, the only non-commissioned officer in his battery to be so honoured. We can imagine him in the dark of night, running from shell-hole to shell-hole with the telephone line running through his fingers, repairing any breakages. It was that dogged persistence in the face of adversity that, ten years on, saw him weather the Great Depression of 1929, just five years after his 1924 purchase of *Glenspean* at postwar boom-time prices.

As a legacy of his soldiering experience he was afraid of nothing. He was straight up and down the line, too, for which he was much respected in the district.

A less desirable legacy was that he had a rather parade-ground manner of dealing with farm staff, expecting similar discipline to that which he had been under in the army. He was a tall amiable man with a quirky sense of humour but judging by my own experience of battling bad temper and shortness in dealing with people it seems that he had that characteristic before I did. Anger would spew up from nowhere without warning. For whatever reason he was always having a 'row' with some employee or other. I can recall just one occasion of his impatience with me in my attempts to wield an axe. Seizing it from me he showed me how it was done. Otherwise my memories of him are all good. I may not have seen much of him during the day because of his busyness with the men: but at lunchtime I remember his grin as he pulled out a matchbox to release a tick – it must have been shearing time. After lunch he sat in his big chair to have a few minutes' shut-eye before returning to work, right on the hour.

We were in the passage when I burst into tears. Dad inquired of me what was the matter and I cried, 'I can't read'; so it was he who instigated my enrolment in the Malvern state school where I did learn to read. I was a real cry-baby. Dad had rented some land between Lal Lal and Yendon. When he bogged the old Reo in a drain there I yelled. He loved fishing. We were in Grandpa's *Bantam*. With the engine running we were pivoting around the snagged anchor, and I yelled. Dad put up with me.

When I was going on twelve, my dad rejoined the army for my/our protection. Before embarkation he bent down and gave me a kiss for which I smiled; it was the only time I can remember him doing so.

Dad was adept at handling small boats with oars or engine but I introduced him to sailing. With John at the helm and me as jib-hand he handled the 'main'. One day, intending to race at Geelong he exclaimed, 'We are going to get a pasting today, boys.' We didn't, but we did have the heavens descend on us with six inches of rain (150 mm).

Dad built a tennis court, its earth surface painted with lines of whiting mixed with hot fat. Tennis parties culminated with Mum pouring tea and the men conversing in the bay window of the sitting room.

Mum and Dad were great entertainers. For the Meredith Golf Club annual championships they opened up the house, taking off internal doors to allow the free flow of guests. The 'playroom' of our childhood, the screened-in verandah, became the ballroom. Tarpaulins hired from the railways provided protection on the outside whilst strips of cloth decorated the inside. Similar arrangements were used for John's and my rights of passage as we grew up. A wooden 'niner' was tapped on the front verandah.

Whether or not his ill-temper inflamed that of my mother I do not know, but in later times they were always shouting at each other, and there were frequent cries from Mum of, 'Go and take a Phenobarb!' (Phenobarbitone was prescribed for both of them to calm things down). Nevertheless Mum and Dad still loved each other and shipped together on a world trip, cut short on its final leg by Dad's death.

(My own endeavour to free myself of crabbiness has had its funny sides. I have frequently had to apologise: apologising is something you are reluctant to do, but like vomiting, you feel better when you have done it.)

Dad's parade-ground manner in dealing with staff contrasts markedly with his grandchildren's positive, yet conciliatory, consultative approach to personal relations.

Following the Great War Dad paired up with his oldest brother Alistair to resume dairy farming at Beveridge. It was there that he met Kay Browne whom he married a year after taking possession of *Glenspean*. (Grandpa Browne, having large estates in Queensland, delayed the wedding because he thought that Dad, being the son of a mere stock agent, was of a lower social class.)

Kathleen Gordon Browne hated her first name because of its Irish connotations, so she liked to be known as Kay. She loved the 'Gordon' because of its (rather dubious) connections to Scottish gentility, and in latter days she would sign herself as K. Gordon Cameron. When grandchildren came along she became known to all as 'Cammy'.

Kay was born at *Narada Downs*, seven miles south-west of Tambo in central west Queensland at the ebb of the nineteenth century. At the age of six she remembers seeing the crumpled figure of her father being carried in from a gig after his horse reared over backwards and crushed him. Whether as a result of the fall itself, or the ride in the gig, Grandpa lost control of the lower half of his body. To reach the railhead at Charleville it took two weeks of travelling in a wagon, with two station hands holding grandfather on a stretcher to ease the bumps. How they coped with his incontinence, I have no idea. From Charleville he was taken progressively to hospital in Brisbane, Sydney and finally Melbourne. With leg-irons Grandpa was able to walk with crutches, swinging his legs through in impressive fashion. On occasion he lifted himself on his hands around the garden to weed it. He bought *Kirkbank* in High Street, Malvern, just down and across the road from the Town Hall. It was there that Kay grew up. Thought to be delicate, Kay initially had a governess, a Miss Davis, who, despite having no teacher training, imbued my mother with a love of literature and of natural history. She loved reading and enjoyed the music of the words for its own sake. Despite her isolation in childhood, when she went to Lauriston Girls School with the Misses Kirkcope, she made many friends. Building on Miss Davis's tutelage, she became dux of Lauriston in her final year. She attended Melbourne University for a year, until her father removed her, deeming university education unladylike, whereupon she took painting lessons and became very adept with watercolours, giving great character to her subjects. During her years at *Kirkbank* between leaving school and marriage she made many additional friends, who became 'aunties' to me. Many were spinsters due to the shortage of men resulting from the Great War's carnage.

At *Glenspean*, whilst rearing us two boys, Kay had to cope with a long passage to the bedroom at one end of the house and the men who ate in the kitchen at the other end. We always ate in genteel fashion in the dining room off the long passage, behind the coloured glass door, which gave us privacy.

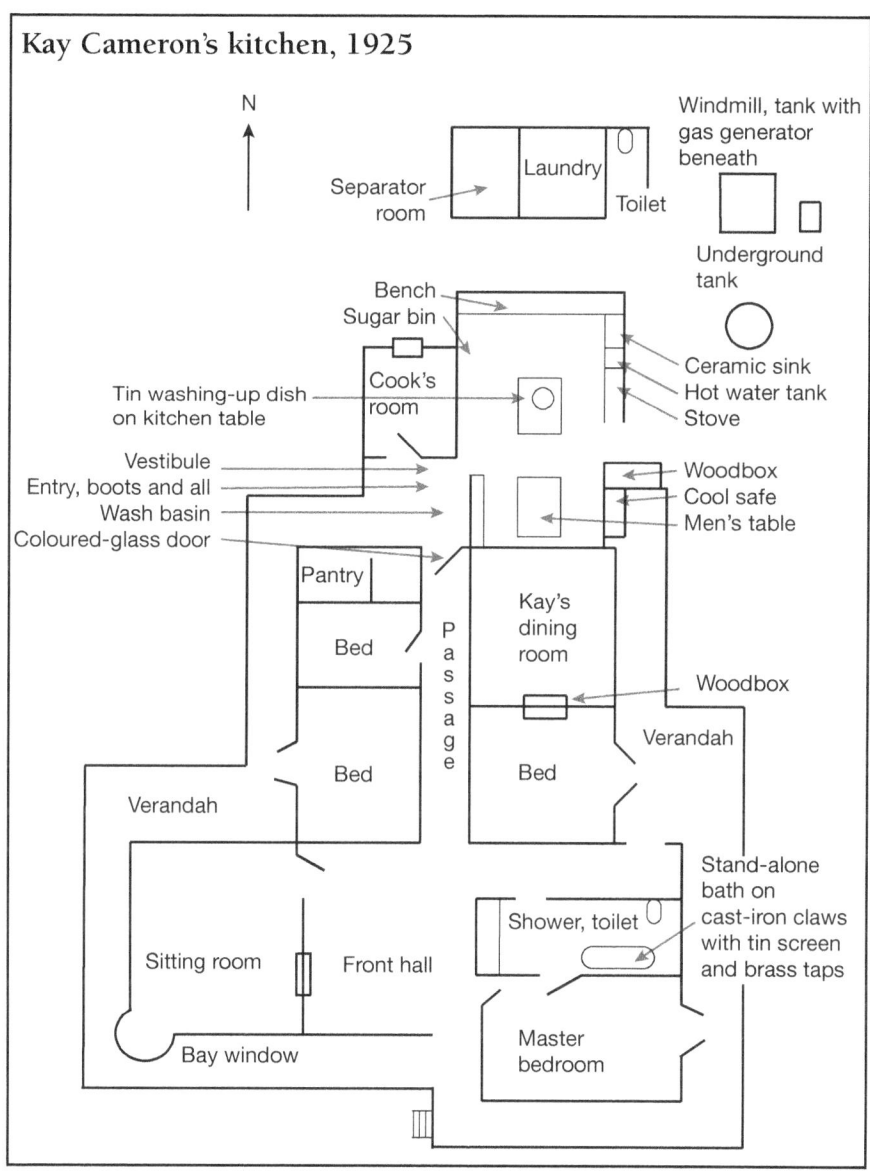

Kay Cameron's kitchen, 1925

When my mother came to *Glenspean* as a young bride in February 1925, cooking was done on the old cast-iron wood stove set into a wide chimney space. The 'cowboy' each day brought into the wood-box a large iron-wheeled barrow loaded with 'foot wood'. A barrow load of 2-foot wood went to the dining room wood-box: both were filled from the outside but accessible from the inside. One of the chores of the cowboy was to bring in armfuls of sugar-gum twigs. Dad would run down the passage in the morning, use a pile of twigs to light the fire in the stove and over its open flames boil a tin kettle for his and Mum's early morning cup of tea. This he claimed to be able to do in five minutes.

The cowboy's main job was to milk several cows. His two bucketsful of milk were carried to the separator room. As the separator was cranked up a bell would ring until it was spinning fast enough; then the milk could be turned on. The cream was added to that from previous milkings, to be preserved by the acidity of going sour. Twice a week a can of cream was consigned by rail to the Wallace butter factory to the east of Ballarat.

In later days the surplus milk from a single cow was set for scalded cream in a basin atop the 'Doover', the coke hot-water heater, coke dust notwithstanding. Either way, butter was made on the kitchen bench from cream by beating it in a wooden churn cranked as fast as you could go. With the buttermilk drained off it was kneaded with grooved butter pats in fresh water, then formed into blocks of approximately 1 lb weight (650 g). Making butter was another job for the cowboy: in later days I did it. (For the chemist: cream is a fat in water emulsion; butter is a water in fat emulsion.)

I cannot tell you anything about Kay's cooking in the early days because I was not there, but I have a memory of Mum plastering a leg of mutton with dripping from previous roastings. Chops would have been grilled over the flames of the stove with the top removed. Potatoes and other vegetables were boiled. Thinking about it, we always had boiled potato, mashed then whipped with a fork, adding a little milk and butter, salt and pepper.

For sweets, custard complemented apples baked with plenty of butter and sugar; or junket accompanied fruits boiled, or preserved using a Fowlers Vacola bottling outfit. It is doubtful if Kay ever baked a cake; but her forte was drop-scones (pikelets). We never went hungry.

Sugar was stored in a large bin opposite the stove. It came in 70 lb (30 kg) bags.

With a cord closing the neck of a sugar bag and attached to one corner of the bag it was slung over the teamster's shoulder to carry his lunch and a screw-top lemonade bottle of cold tea.

In my memory George Scarcie rode a pony to where the team was stabled close to the paddock which was being worked up for sowing down to improved pasture.

The kitchen had a ceramic wash-up basin set into a wooden bench-top in an awkward corner: so washing up of the dinnertime dishes was done in a 'tin' basin on the kitchen table with a flat 'tin' drainage pan. After dinner at night Dad would call out, 'Come on boys: up and at 'em'. It was our job to do the drying up and putting away. Apart from that I can say that Dad did not produce much innovation on the kitchen front: that was up to Kay.

Kay always liked to do things nicely. The lunch table was graced with little blue dishes of butter-balls made by rolling small portions with the butter pats. After lunch the butter, and jams too, were put away in the safe and the dishes washed – and the jam spoons licked by me. Those goodies were inconveniently stored as far away as possible in a ventilated and shaded 'safe' on the east side of the house. It was equipped with a vertical sliding shutter which was jammed open. Evaporative cooling was by a cloth draped over the food and into a dish of water. Other people had safes of fly-wire or perforated metal which could be hung in a shady place, taking advantage of any breeze for evaporative cooling.

Meat was kept in a stand-alone fly-wire safe shaded by the highest of the eight old pine trees to the west of the house. The nearby chopping block comprised a length of tree trunk set into the ground for stability.

Refrigeration

In the year before the invention of the Frigidaire system, which is still basic to modern gas refrigerators such as those used on camping trips, Dad bought Mum a Hallstrom Refrigerator, the workings of which had to be heated by kerosene burners for two hours every morning, driving the refrigerant gases through a cooling tank, to be stored as liquid in small cylinders set into the refrigerator cabinet. With the heating completed, volatilisation of the liquid refrigerant cooled those cylinders and the cabinet. There were slots which could carry a small tray of ice or ice cream. Long after it had been superseded as a refrigerator the Hallstrom cabinet was a very convenient storage for party drinks.

Upon discharge from the army at the end of 1943 Dad was able to buy a four-oven AGA 'Heat Storage Cooker' thereby saving the labour of cutting wood. It burned coke, a by-product of the Geelong gasworks. There was a separate coke water heater, 'the Doover'. Thereafter coke dust was ever present until the Doover was made redundant by mains electricity in 1958. The AGA stove was converted to oil in 1967. With it, long after Dad's time, I was able to cook the shearers a hot midday dinner whilst at the same time classing the clip in the shearing shed. The roast spent 40 minutes in the roasting oven. It was transferred to the simmering oven before shearing commenced at 7:30 a.m. At the 11 o'clock smoko I ran down and placed veggies in the baking oven.

In winter, coming in cold from the farm, I would sit on the warming plate of the AGA and listen to June as she prepared our dinner on her return from her day at Shannon Park.

Electricity supply

From 1946 kerosene lamps and candles were replaced with electric light initially provided by a wartime disposals 32-volt generator and a bank of batteries. 'Herbert' had to be cranked vigorously to get him to start. From 1951 we had a 110-volt system sufficient for us, two staff houses, the shearing shed and a large commercial type refrigerator in the 'rat-house'. The generator was driven by a diesel engine massively heavy for its 7 hp. Made by Ronaldson & Tippett of Ballarat, their motto was: 'We don't know how strong the metal is; so we use plenty of it'. Needless to say the company did not survive long in the innovative postwar years. The engine was water-cooled with a large cooling tank outside the shed. 'Semi-automatic', it was supposed to start when a certain load came on. We had to turn on all the lights we could think of to prompt it to start. Frequently there was not enough charge in the batteries so that it had to be cranked by hand whilst holding down the exhaust valve until it was spinning fast enough to fire. When the State Electricity Commission (SEC) power came through in 1958 we were pleased to palm off 'Ronnie' onto Ian Sutherland who was outside of the group scheme which brought it.

Communication

Three times a week Mr Henderson in his horse-drawn buggy would deliver mail to our mailbox at the crossroads five miles west of Meredith. He then turned south on his way to the Bamganie post office cum telephone exchange. With the mail would come copies of the Melbourne *Argus* and a weekly paper, long since gone to oblivion, which in our childhood John and I would rush at to read the latest adventures of Brick Bradford. Mum and Dad received by rail a monthly parcel of library books from the Melbourne Athenaeum lending library.

We always had a telephone but the Postmaster General's Department service from Meredith only came out five miles as far as the crossroads before diverting southwards towards the Bamganie telephone exchange. It comprised a pair of wires for each subscriber. The wires were supported by white insulators on crossbars carried by telephone poles every so many yards apart.

From the crossroads it was Dad's responsibility to make a connection to our house, which he did with a single-wire earth-return system. It necessitated having a connection from one of the two wires to earth at the crossroads and another 'earth' at the point where the phone line came into the house. The 'earths' comprised some small diameter galvanised pipe driven into the ground. In order to make a good connection those two earths had to be wet with water. The single wire was not copper, but more robust ten gauge (2.5 mm) galvanised fencing wire.

Theoretically any joints in the wire along the line would have been soldered. To heat the soldering iron entailed using a kerosene pump-up blow-lamp, itself preheated with a little petrol in a small collar designed for the purpose. The soldering 'iron' of copper was heavy to keep its heat during the actual soldering process. It and the joint to be soldered had to be 'tinned' after preparation with 'killed' spirits of salts (hydrochloric acid neutralised with little bits of zinc or failing that scraps of galvanised roofing iron with their zinc coating). It was obviously a job for Dad's handy-man, Emmett, who would have chosen a day without wind so as not to blow out the lamp or cool the soldering iron.

Connections were made from the aerial and the earth to the wall phone in the dining room. To make an outside call one cranked the handle which rang a bell in Mr Wicking's Meredith post office. During his 'open' hours he would then make connections to the rest of the world, passing the request for a particular number down the line to the next telephone exchange; so when you were wanting to ring the manager on Grandpa's farm at Exford, overlooking the Melton Dam, Mr Wicking would pass the request on to the Geelong telephone exchange from which it would progress to the main Melbourne exchange and then out to the Exford telephone exchange which would then ring the incumbent on Grandpa Browne's Nether Place. 'Is that you Spedding?' Dad would bawl into the mouthpiece on the wall phone in a voice loud enough for Spedding to hear without it.

Water supply and sewerage

Kay had hot water available at the kitchen sink, at a hand basin in the vestibule and in the bathroom at the far end of the house.

The bathroom had a flushing toilet – a water closet or 'WC'. Dad, Neil Wilson Cameron, had apparently been mercilessly teased at school about his initials NWC. When our children came along and were needing to be named we had to be careful about incorporating the name William for their maternal grandfather, William Upjohn.

There was another toilet in the outhouse which as a youngster I would use after making a careful search to ensure there were no snakes. There never were any, but I was dead scared of snakes, even as an idea.

Water was heated by passing through copper pipes set into the wood stove. It was stored by way of a thermo-syphon in a hot-water cylinder adjacent to and above the stove so that cool water gravitated from the bottom of it into the stove and less dense hot water would rise up to the top of the cylinder. Because of the low pressures generated from the overhead tank supplying water to the house, hot water was carried to Mum and Dad's bathroom in ridiculously large 1 inch outside diameter copper pipe. So much water was lost that you thought twice before running it until hot.

Roof water was collected in a 5000 gallon galvanised iron tank, or, depending upon location of downpipes from the spouting of the rambling house, via earthenware pipes to the underground tank. As it did so it passed over the earthenware pipeline from the toilets to the septic tank so that one had to be careful that the grouting of one length of pipe into another was sound. Each length of 4 inch internal diameter glazed ceramic pipe used for both lines was 3 feet or less in length and the cement–sand mortar of the grout was rigid so that any movement of the soil could crack it causing a leak. Bends ranging from 15 to 85° were used as necessary.

The sewerage was poor, there being no satisfactory solution available to Dad. From the outside toilet and from the bathroom lines were conjoined in a pit located under Kay's rose bed and then gravitated on a very slight gradient to the distant septic tank. The roots of some cypress trees invaded any cracks in the pipeline. It was always causing trouble. There were numerous pits with openings to the top of the pipeline, enabling a wire to be inserted to clear any blockage. The outlet from the septic tank also had be kept clear of invading roots.

Kay did ride, but generally took a rather remote interest in the running of the farm and instead developed a large garden, despite having only limited water from the house dam.

Then with the onset of the Second World War most of the farm staff hastened to join the fray before it was all over. After the fall of Dunkirk, appalled that all the good work which the 'Diggers' had done in France was being undone by Hitler, Dad also rejoined the army. Mum managed the farm in consultation with Dad's faithful helper Emmett Cavey over an evening glass of sherry.

Emmett Cavey

After Dad's return from being overseas for eighteen months, Mum apparently found being a housewife unsatisfying and frequently retired to bed with a migraine headache. They disappeared when the Country Women's Association gave purpose to her life. She founded the Meredith branch, graduating from it to the Geelong Group of branches, then to the state and then national association, taking every office available except that of secretary. She joined the Associated Country Women of the World, and led the Pan Pacific Women's Association. She travelled widely on the profits from Queensland. Once again she made many friends in the Pacific region including Queen Salote of Tonga, others in the Philippines, Indonesia, Burma and also Canada.

Upon my marriage to June it was found that there could not be two women in charge of the house, so Kay retired to live at *Pine Cottage* Sorrento to entertain us and grandchildren over summer holidays, when she would direct operations after breakfasting in bed. June, or sister-in-law Beb, would go to enquire of her the cuisine of the day, one or another of the children having already climbed into bed with her to be read an early morning story.

Cammy was frequently called on to mind the children by taking them to the beach. On one such occasion the children were nearly orphaned. June, Derek Colvin and Ian Haskins accompanied me on a trip to Mud

Island. After walking around it for an hour we set sail for Portsea with a light northerly drifting us along and June saying, 'This is the sort of sailing I like.' I instructed the crew on how to set the spinnaker but had no sooner done so when I saw a dark line on the water in the direction of Port Phillip heads.

'Down spinnaker,' I cried.

'C'mon Skipper, we've only just put it up.' By the time we had doused the spinnaker I could see white water between us and Portsea. 'Down main!' It was only half down when the cold front hit us, driving us towards Frankston. With only the jib set I sailed as close to the wind as I could, with the wind on our starboard beam. After an hour we made land on Port Phillip's southern shore at Tootgarook. Fortunately we had some silver coins in the drawer and were able to ring a relieved Cammy and ask her to come and get us.

(A point of nautical interest: the distance from where the gale struck us to Tootgarook is eight nautical miles; so under jib alone we were making 8 knots, whereas according to Uffa Fox's formula, *Mañana's* maximum speed on the wind under full sail was 5.6 knots. I guessed the wind speed to be 50 knots.)

June's only thought during that hour was, 'My poor orphan children.'

We made Cammy welcome on her frequent visits to us at *Glenspean* when she was in demand to do the flowers but found difficulty washing up for us at the sink because of her short stature. She often brought us visitors representing her environmental and international interests. Most definitely a monarchist, at the time of the Royal visit in 1956 she brought us as house-guests several sailors from the Royal yacht. Many others whom we entertained were Cammy's friends gained from membership of the Associated Country Women of the World.

Cammy spent a lot of time staying at the CWA (Country Women's Association) club in Lansell Road, Toorak. When she was crossing Toorak road to post a letter she was run over by a car, thereafter spending six months in hospital. Now Cammy liked things just right. She had her own recipe for 'Italian' salad-dressing, with spices augmenting the oil and vinegar. Her special nurse put it in the hospital refrigerator labelling it *Mrs Cameron's embrocation*. In her dotage we took her home to live with us but in the end, Cammy having had a stroke, the demands were too great. We had to give up and send her to nursing homes, of which the less said the better. We tried to visit her at mealtimes to assist her to eat, otherwise the hospital staff would feed her whilst yacking to each other across the room. In the end they woke us up in the middle of the night to say that she had died, the only time that they had given her personalised attention.

June planted a memorial tree for her in the grounds of the Meredith Uniting Church.

Neil and Kay on their wedding day

3 Neil's and Kay's antecedents

Alexander Clayhills Cameron

Through Alexander Cameron's enterprise our branch of the Cameron clan came to Australia, even though he overreached himself, ending in financial disaster.

Alexander's career may be likened to a broken shoot of a precious vine being found and lovingly potted up by a gardener. The rooted cutting was transplanted to fresh soil in which it continued to thrive. It is the stock which, when cross-fertilised year after year, continues to bring forth fresh fruit in a great variety of flavours, some sweet, many strong.

Alexander was born on 19 February 1823 in the Parish of Balgarthno, Forfarshire, Scotland, the youngest child by four years of William Cameron, born 15 April 1762, and Elspeth Wilson, born 12 January 1780. They had married on 2 August 1802. They had seven children of whom, apart from Alexander, only three survived to have children of their own.

William and Elspeth were lowlanders, tenant farmers in Fife. They both died and the orphaned Alexander was taken into the Wilson family, that of Elspeth's brother Henry Wilson. The Wilsons found Alexander to be a likely lad and gave him a good education before apprenticing him to a flour miller.

Enter into the story Phillip Russell of Van Diemen's Land. He came to Scotland recruiting labour for his Tasmanian farm. In an indenture dated 10 August 1842 he offered young Alexander a three-year contract as ploughman and general hand at £20 per annum in the first year, £22 in the second and £24 in the third, all 'found'. The indenture said that Alexander was to take ship to Van Diemen's Land, which means he would have arrived there a bit before his twentieth birthday. When

Phillip Russell died, it seems after a couple of years Alexander took other employment. Alex Reid, farmer of Rathno, signed a certificate dated 25 April 1845 saying that Alexander had been in his employment for upwards of 12 months.

Leaving Reid, Alexander decided to cross Bass Strait to the mainland. He arrived at Port Welshpool and gained employment as a storekeeper for Angus McMillan, the man who had opened up Gippsland to white settlement.

Phillip's brother George had his eye on the young Cameron. When the Tasmanians began to look across Bass Strait to the wide open spaces in Victoria, George Russell head-hunted him to manage the Clyde Company's Terinallum Station which he did for 11 years from 1846 to 1857.

During that time there came the gold rush and much of the fourth volume of the Clyde Company Papers is taken up with Alexander's letters to George Russell explaining his labour problems as his workforce headed for the goldfields.

From employment with George Russell, Alexander was enabled to buy into a partnership which owned *Kongbool* near Balmoral in far Western Victoria. The Scottish waif became a Western District squatter.

Kongbool was next door to both *Satimer*, which he eventually bought, and to *Englefield*, owned by Duncan Robertson whose eldest daughter Jesse he married at Portland on 29 December 1858.

Duncan was a true Highlander: his wife Margaret Stuart never did learn to speak anything but Gaelic.

Alexander and Jesse had two surviving children who were born at *Kongbool*: Wilson born 1861 and Jesse born 1865.

From being a partner in the ownership of *Kongbool*, Alexander lashed out and bought *Satimer*, putting down a small deposit and incurring a large overdraft. Then his apple went pear-shaped. To put a curb on the areas of land that squatters could own, the government offered each a square mile of his property as a pre-emptive right at the same time making the remainder of his property open to selection. To retain *Satimer*, Alexander had to buy out the selectors, increasing his indebtedness. At about the same time, interest rates were running at 8 per cent and there was a drop in the price of wool. There was no market for meat from which Alexander might have augmented his income. Surplus and aged sheep were just boiled down for their tallow, which was shipped to England.

When Alexander died on 12 January 1874 he left a huge debt so that *Satimer* had to be sold.

With little or no capital, his son Wilson Cameron sought work as a stock agent at Victoria's Newmarket saleyards, joining the firm of Pearson Rowe Smith. Wilson did sufficiently well that he was able to buy a large block of land at Heidelberg, north-east of Melbourne. The story is that Wilson took possession on 31 August and using second-hand materials, possibly with monetary assistance from Charlott'e father, had built on it a large house by the end of November. He named it *Balgartnie*. Just how much physical labour he put into building it is unknown but the work was supervised by a Henry Wilson. It was pretty smart work but then people lucky enough to have a job worked long hours on a six-day working week. At some stage Wilson was also able to buy *High Park*, a small farm on the Kilmore–Lancefield Road.

So that is how the scion of the Western District squatter was downgraded in social stature to being a mere stock agent, something of which my maternal grandfather John Browne was aware when considering the marriage of my father Neil Wilson Cameron to his daughter Kay.

Wilson married Charlotte Walker with whom he had six sons. My poor Granny. Dad had stories of the boys playing cricket up and down the hallway.

This brings us to the story of Granny's father, Joseph Henry Walker, another of my entrepreneurial great-grandfathers.

As related by Uncle Malcolm, Joseph was working in an architect's office in London when a friend came in asking, 'Why don't you go to Australia?' Questioned as to how he would get there, the friend replied that there was a ship going to Australia in a week's time and they needed a tutor for the migrants. So Joseph bade goodbye to his mother and took ship to Australia. To the literate he taught arithmetic and algebra. The others he taught to read. Joseph's diary, written up by Granny in 1902, shows that it was part of a general migration program.

They landed at Port Adelaide and were faced with the dusty seven mile track to Adelaide. In winter the track became a bog. Discovering a shortage of accommodation Joseph decided to build some houses, so he bought the between-decks accommodation from the ship's captain who had to make room for a return cargo of wheat or copper ore. To retrieve the materials he threw them out the stern window and swam them towards shore, only to get marooned on a sand bank. He built a couple of wattle-and-daub houses with shingle roofs using the ships fittings for the doors and windows. He was able to let them out at 25 shillings a week.

With news of gold discoveries at Bendigo he decided to join others in the rush there. The party had a number of heavily laden drays and the horses to pull them, with others to ride. The trip involved crossing a desert, then swimming the Murray River. Instead of the expected fortnight it took six weeks by which time there remained of the party only Joseph, a friend and two ladies plus just one small horse to pull the dray.

'Golden Gully was a site never to be forgotten – a long valley with tents on each side, men scattered all about, either sinking or washing with wooden crates; no time for larking or quarrelling; each one eager to get to the bottom of his hole about 12 feet deep with gold at the bottom. You haul up gravel, wash it in the cradle; then take the sand from the bottom, put in a tin dish, twirl it around and there is the gold with some times a nugget which makes you smile.'

Joseph decided that not being used to manual work he could do better by contracting. He built a hospital which unfortunately threatened to fall down until the supports for the heavy roof were hastily strengthened. Underneath the sagging roof was little Dr Hunt calming patients.

Water was scarce. Joseph noticed an abandoned shaft full of sweet clear water; he rigged a Californian pump and made two water carts out of barrels. He paid men three farthings (3/4d) a gallon for delivery of the water and contracted to supply it at tuppence three farthings (2 3/4d).

Hearing of an unused quartz crusher lying on the wharf in Melbourne, he had it sent up.

He was riding home after work when there was an awful blast as the boiler blew up. Bendigo's first battery! It cost him £4000.

He had married Charlotte Hawes, one of the pretty girls he had brought from Adelaide. He took her and son Harry on a trip to England to calm him down and catch up with family. Their daughter, also Charlotte, married Wilson Cameron.

The spread of the Cameron vine

Here then is the genealogy of the Camerons of *Glenspean*: five Australian generations showing initiative and innovation.

Alexander Clayhills Cameron had three children, including one son.

Wilson Cameron married **Charlotte Walker** and had six sons.

Neil Wilson Cameron married **Kay Browne**. They had two sons, John Gordon Cameron and **Neil Gordon Cameron** who married **June Upjohn**.

Their eldest son **Alexander William Neil Cameron (Sandy)** and wife **Julie Brown** are the spirit upholding the Meredith Dairy

Of Wilson's six sons, the four oldest saw service in the Great War.
1. Alasdair farmed north of Melbourne.
2. Colin was killed at the Nek at age 23, having been recently promoted Sergeant Major; 'much loved and a fine horseman', also presumably a churchgoer because he was a member of the board of management of the local Presbyterian Church.
3. Neil Wilson Cameron.
4. Evan, who had a farm, *Challicum Hills*, at Buangor, then a retirement farm of 500 acres just west of Ararat, which he proceeded to improve. He recommenced the triennial Cameron reunions, Granny Cameron having had one at Wattle Park in 1939, when Dad took off his waistcoat playing cricket and someone filched his gold pocket watch.
5. Norman and his wife Nancy lived in a covered wagon when they first went on to a dry-land farm at Carrathool near Hay, New South Wales. Subsequently they were 'made' when they began irrigating with drainage waters from the rice fields. Their daughter and three sons all became farmers. Numbers three and four went into the big time, but over-extended themselves and came unstuck.
6. By trading in livestock Malcolm also went into the big time and stayed there. He had farms in the vicinity of Euroa and in the Riverina.

Kay Browne's wedding veil, whereon there hangs a tale

Samuel Brown, youngest son of Alexander Brown and Marion Gordon of Scotland, wanted to marry an Irish girl, Eliza. His family demurred; so he up and took his girl to Australia where they were married in the Royal Hotel, George Street, Sydney on 25 May 1864. He then moved to Queensland where he already had an older sister, Marion, married to a John Beck.

Sam's older brother John said, 'If you are going to Australia I am going too!' But for reasons unknown he chose to go to Victoria.

On 15 March 1840 John Brown arrived in Australia on the ship *Indus* which also carried Miss Ann Drysdale and friend Miss Caroline Elizabeth Newcomb. Ann had been a sheep farmer on property she owned in Scotland, but decided to emigrate to Australia for reasons of health. On arrival, being of the moneyed class, she had useful connections which enabled her to gain Crown leases; then in 1843 she purchased the freehold of *Coryule* on the Bellarine Peninsula.

John Brown in partnership with Thomas Sproat depastured sheep over half the Bellarine Peninsula, but with less legitimacy than Ann Drysdale who ran her sheep on the other half. When it was becoming 'too civilised' he removed to a properly named *Moppianimum* near Scarsdale where in 1849 he purchased the share of Thomas Sproat. On 17 July 1852 he married Eliza Tennant Linton. An unsigned note suggests that her father, Joseph Linton, was killed when thrown from a buggy as the horse shied during a thunderstorm. The property was on auriferous country and, come the gold rush of 1851, the diggers dug him out: so John Brown sold *Moppianimum* in 1859 and moved to *Narada*, Anakie, 20 miles east of Geelong. It was not a good move because *Narada* was in the rain shadow of both the Otway ranges and the You Yangs. Neither was it a good move to build a substantial bluestone house thereon. He over-extended himself and had to sell *Narada*, retaining only the back paddock under the shadow of the Brisbane ranges, which he named *Narada West*. I have memories of going there with Mum and Dad to visit John Brown's two children who never married — Aunt Marion and Uncle Bill. A nervous child, when in the old Reo car we were chugging back up the fault line to top the Brisbane ranges, I was always afraid that we would not make it.

The Reo was Grandpa Browne's wedding present to Mum and Dad. It soon doubled as a farm car. One time there was a calf to be sold in the backseat which ate Mum's hat. Another time, with no brakes, Mum managed to knock down a verandah post in Moorabool Street Geelong. The Reo ended up cut down to a flat-top ute with petrol gravity-fed from a tank on the tin roof. One pull on the starting handle was sufficient to get it going. Before the age of the four-wheel-drive but with 20 inch wheels, it was almost as good as one, able to go almost anywhere on the farm.

When things got tough for him, forcing him to sell *Narada*, John Brown's four older boys heard from their cousins in Queensland that land was being thrown open for selection in the west, so they took ship to Brisbane and initially settled on *Narada Downs*, seven miles south-west of Tambo, 500 miles west of Brisbane. To it they added a separate property, *Toliness*, further south in the direction of Charleville. One of the brothers, Montgomery, went off on his own. A fourth brother, Malcolm, after an argument, was sent home by the others never to be spoken of, nor spoken to again. 'Forgiveness' was not in the Browns' lexicon. To distance themselves from him the other brothers added an 'e' to their name — Browne.

That left two brothers, John Gordon Browne the oldest and Thomas Maitland Browne the youngest of the four, one for each of the Tambo properties. When after 15 years John and Tom came south looking for wives they found two of the Nicol sisters only too willing to oblige. Their father, James Nicol, had migrated to Geelong with the London bank where he became its manager. My grandmother could remember standing on the balcony of the very handsome bank building on the south side of Malop Street opposite Johnson Park. When he died the family retired to *Kinloss*, a small house in Aphrasia Street, Newtown, where the several daughters had been squabbling over the much reduced family income. The young men took their new wives home to the outback. It was a marriage of convenience for all parties.

I believe my mother to have been born at *Narada Downs*, handy to Tambo, where her mother was staying with her sister Frances, married to Tom Brown; I think that *Toliness* was home to Mother and her parents John and Annie Brown. It was where grandfather had his accident six years later.

Moving forward a generation, as a paraplegic my grandfather John Gordon Browne had loaned Mr Carr-Boyd money to prospect for gold in Australia's Kimberley region. If he had been prospecting for diamonds the story might have been different, for the Carr-Boyd ranges are now the site of Australia's Argyle diamond mine. Anyway, unable to repay John Browne, when the latter's daughter Kathleen (Kay) became engaged to my father, also Neil, he had his daughters crochet, embroider, or whatever you have to do to make a veil of Convent Limerick lace.

Grandfather was a sitting duck for sharks (sorry about the mixed metaphor) wanting a bite of his wealth earned during his twenty-one-year venture into western Queensland with his two brothers who were close to each other. They had prospered. After 21 years my grandfather had his fall with a horse, becoming a paraplegic.

Although he could walk with leg-irons and crutches and on occasion would sit on the ground gardening, lifting himself around on his hands, he spent much of his time sitting on the verandah with his feet up, in front of his big house in High Street, Malvern, Victoria. The only time that Grandpa spoke to my brother John and me was when we were playing on the tiled verandah; he told us not to make so much noise. At dinner he sat at the head of the table to carve the roast and on Sundays would also come into the dining room at 11.00 a.m. to listen to the church service on the wireless. What stories he could have told us. I suppose that he had spent too much of his time alone. There was not much conversation between him and Granny who lived up a narrow flight of stairs at the back of the house.

4 Neil Wilson Cameron's innovative thinking

Glenspean was covered in rocks and tussocks. In the early days, ploughing with a stump-jump disc plough loosened some of the rocks which were then lifted onto a sledge for removal. Everything was done by manual labour with motive power provided by horses, of course.

According to Dad he took possession on 2 February 1924, and on 3 February he put the plough into the nearest sizeable area, one bounded on the east by the unmade road which runs close to the front gate where the formed and gravelled section terminated. The remainder of the road was a rough dirt track bounded on the far side by a gorse hedge belonging to Dad's neighbour, Alf Wells. Post Second World War, and presumably with the compliance of Mr Wells, Dad got permission from the Bannockburn Shire to close the unmade section of road so that he could spray the gorse with the only herbicide available — arsenic. Its hedge was home to rabbits and driving down the road you could see thirty rabbits run across to its shelter.

The paddock that Dad ploughed so long ago now comprises the five Ryegrass paddocks, so named for what he sowed.

'I sowed 2 pounds of sub-clover seed/acre with superphosphate, and with it 2 pounds of ryegrass so as to be there when fertility improved.'

He sowed the only varieties available, Mount Barker sub, and Victorian perennial ryegrass, both inferior by modern standards. That was after taking off a couple of crops to finance the exercise and to feed the horses. Cropping removed competition for new pasture species from the former unproductive pasture of wallaby grass and weeds, which was only capable of carrying in Dad's words 'two wethers to 3 acres — badly'. The new pasture with exotic species was capable of sustaining four wethers per acre, a sixfold increase in productivity.

Pasture improvement was Dad's prime initiative – one not followed by the district in general until twenty years later, following The Second World War and the postwar influx of soldier settlers.

Shelter belts

Despite his shortage of funds, Dad planted one thousand trees every year in three-row shelter belts, using three rows of radiata pine (*Pinus radiata*) planted in winter then as now, open rooted and suitably fenced off for protection against stock. Just how Dad financed the requisite fencing I do not know. Once again tree planting initiative was not generally followed in the district for another twenty years. *Glenspean* as he bought it was by no means devoid of trees. There were remnant drooping she-oaks (*Casuarina stricta* in old parlance) part of a forest unique in this higher rainfall district and about 10 miles by 5 miles in extent, which one feels must have never been burnt, for unlike eucalypts and acacias, casuarinas do not regenerate well after fire.

There was an existing sugar-gum plantation *(Eucalyptus cladocalyx)* providing some protection to the sheep yards from the prevailing winter north-westerlies, and to the west of the house there were eight big old pine trees. There was, and there remains, a line of mahogany gums (*E. botryoides*) along the roadside. There was also a sugar-gum plantation on the south-west boundary of the front paddock, the 20 Acre. When the roar of the wind in those sugar-gums heralded a squall from the south-west, the sheep in the front paddock would go running for shelter behind the large plantation which, early on, Dad had fenced off to the south-east of the house. He had done so to conserve some remnant drooping she-oaks.

Not short of something to do Dad also constructed a bare earth tennis court, the venue for many a tennis party with other members of the minor squattocracy to which we belonged. The real squatters were those squatted on large tracts of land in the early days of Port Phillip settlement. By virtue of having a private school education our group included some who were only smallholders. Dad also planted an orchard, on raised beds to combat the prevailing winter waterlogging of our duplex soils overlying basalt, but the tennis court and orchard could hardly be labelled as 'district initiatives'.

Dad's tree planting initiative did, however, extend into the community. When Lex McNaughton provided both the site and the building for a new Woodburn school located at the crossroads, Dad with his faithful helper Emmett Cavey planted a pine shelter belt on its north and west sides. The same year, 1936, saw the birth of the Meredith Golf Club on the racecourse reserve. Dad with Henry Bolte and other members planted a couple of pine shelter belts. One, recently demolished, was triangular to provide an angle in one of the fairways. The soil did not allow for the digging of bunkers which would have become full of water in winter, so to provide hazards in the 'rough', he and Henry dug up and replanted tussocks from our creek flats.

To provide shelter for travellers on the station platform Dad formed a short-lived Progress Association whose members planted a pine shelter belt on the west side of the railway reserve.

Cattle

Once again Dad led the district in farming practice. Early records show that Dad had put on 200 head of dry cattle to eat the tussocks on paddocks not yet subject to pasture improvement. Thereafter Dad always had cattle. He started a herd of cows and calves. Uniquely for the district, they were black – Aberdeen Angus.

He departed from district practice, bringing the calving date earlier from spring to winter, then to autumn so that the calves were old enough to benefit from the surplus spring growth of Dad's improved pastures. They could then be sold as sizeable vealers off the mother by the New Year, the end of the green-feed season in those days. By 1939 most paddocks had been sown to improved pasture. His strategy was sufficiently successful that he was able to complement his herd of cows and calves with an Angus stud. Then the War intervened.

Sheep

Dad's inaugural flock of fine wool Merino ewes produced seventy per cent of lambs at marking but by the following spring only thirty per cent remained alive, the others having succumbed to a combination of poor nutrition on the native pasture and worm disease. So he resorted to buying in crossbred ewes for fat-lamb production, complemented by Merino wethers, also bought in.

The only treatment for worms was bluestone and nicotine. The nicotine was to kill intestinal worms and the bluestone, copper sulphate, closed the oesophageal groove allowing the nicotine to bypass the first three stomachs into the rumen, the fourth or true stomach. It was administered by a tapered drenching tube with a hole in the side so that when it was dipped into the Blu-nic solution and withdrawn it held a fluid ounce of the drench. With a finger over the hole it was poured down the sheep's throat.

Finance

Dad paid a postwar boom price of 10 guineas ($21) an acre for *Glenspean* with only a 20 per cent equity. When the Great Depression came in 1929, Dad was up against it. In buying *Glenspean* he had refused help from Kay's father, but by this time when he would have accepted help, Grandpa Browne had been robbed of most of his money by sharks who had gobbled it up on projects that were half a century ahead of their time.

One was Gippsland Oil, which they tried to find *on land*. Many years later it was found under the sea in Bass Strait. Grandpa's youngest brother, bachelor uncle Tom Browne, was able to lend him some money. Besides that he was mortgaged to the fingernails. He went to his wool brokers for a loan to pay for superphosphate but was refused. He went to another and the manager, Harold Anderson I think, said, 'Double it!' Dad was burdened with debt until the wool boom of 1951 occasioned by the Korean War and the need for woollen uniforms for the American soldiers. I remember him sitting in the big armchair by the fire in the old sitting room and deciding that he was earning threepence (pronounced thrippence, at the time the price of a large ice-cream cone) every second — or was it every minute?

To take on a property covered with rocks and tussocks was a test of Dad's determination: to hang on through the Depression was even greater. He was aided by his resilience of thinking. He continued with pasture improvement of paddocks year by year. It was not emulated by anyone else in the district until post Second World War. It was then given a kick along with the arrival of the soldier settlers in 1952 whose leases were conditional upon them improving their pastures.

By the same token he abandoned the district practice of autumn lambing for a winter – early spring lambing. To cope with losses of lambs due to cold stress in 1944–45 he built the 'Jesse' lambing shed (named after the Jesse MacPherson maternity hospital). He did not then know, nor did anyone else, that the best protection against cold stress in newborn lambs is to have the mothers very well fed. Once again, he had problems with pregnancy toxaemia, the cause of which was not then known. His solution, wrong as it turned out, was to send the fat ewes on a route march around the road towards the middle section of the farm, making them walk back the next day.

Reflecting on Dad's life, it was he who in 1948 had the foresight to establish Glenspean Partners, of which, at the age of twenty, I became then a very junior partner.

By 1952, the year of Dad's untimely death, his farm was still recovering from the privations of the war and its aftermath. Dad's ten-month world trip that year with my mother meant that I had of necessity already taken charge of my end of the farm. Then after walking with Mum on the Tasman glacier on Mount Cook New Zealand he died of a massive coronary. The 'Boss' as he was called was much lamented.

Dad taking Grandpa Browne fishing in the *Bantam*, c. 1933, Grandpa dressed for the occasion, with a board on his knee for filleting flathead.

Mum and Dad on the Tasman Glacier, Mount Cook, hours before he died.

PART TWO

5 Neil Gordon Cameron: early days, growing to manhood after a shaky start

Of my early days probably the less said the better, although I have fond memories of playing with my older brother John in the large garden, or dog-paddling across a corner of the house dam as he swam its length.

We had numbered hideouts in the large garden and around the farm. From whom or from what we were hiding is unclear. Perhaps it was the initiative to get away from it all, to be just us. One hideout was under a 'New Zealand' strawberry bush, which at the bottom of Julie's garden in 2018 is reaching the end of its days. Another hideout was under a pile of new fence posts that had been thrown off a truck in a higgledy-piggledy fashion, leaving spaces into which we could wriggle.

We had a governess, Miss Clarence – *Clarrie*, who taught us little. We sat at an improvised table comprising an old wash stand. Using a draw knife, Emmett had carved away part of the rail supporting the top to make a space for our legs as we sat on stools.

Woodburn Creek State School

The school reopened at its crossroads location in third term 1935, Mr Lex McNaughton having provided both the land and the building complete with fireplace. Dad made the opening speech during which two distracted motorists collided at the crossroads. One shaken up lady driver was asked in for a cup of tea. Mr McNaughton's fireplace was an essential part of school life for two generations. School was protected from the prevailing winter north-westerlies by pine shelter belts planted by Dad and Emmett on the north and west sides.

We attended Woodburn in 1936, riding our ponies the 2.5 miles to school, John leading on Darky and me, when I was not sick, following half a mile behind on Topsy.

John was two and a half years older than me. In childhood I never had a playmate of my own age: there just were no boys in the district.

In 1937 John went to boarding school, leaving me with no playmates at all. Mum used to drive me to school and I would walk home accompanied by the Nolan girls, Mary and Nelly.

There was a portrait of Joan of Arc on the wall of the school room, and head teacher Alf Hailey taught us to sing 'The Marsellaise'. On Friday afternoons he would lead the whole 13 of us across the paddock to his small house owned by Mr Barnett (now *Lochaber* goat farm) for a musical afternoon, banging drums, et cetera. I remember coaching Lorna McNaughton on the alphabet — Mr A says 'a'; Mr B says 'b'. We learnt to write sloping pothooks and jayhooks in our copy books but not to form letters nor to make them into words.

Park Street

After three years at Woodburn Creek and going on age ten I still could not read so they sent me to stay with my grandmother in Melbourne's outer eastern suburb of Malvern where I attended the Park Street junior section of the local state school. Its asphalt playgrounds were so different from Woodburn Creek with its grass and trees. At playtime there was a strict segregation of the boys from the girls.

At Park Street there were plenty of boys, one of whom befriended me. We walked home together; but as he went on past my grandma's house I never thought of asking him in. The years of solitude at Woodburn Creek had left their mark, one which remains to this day. I have difficulty making myself share my thoughts with others. Writing *The Camerons of Glenspean* is basically a solo effort although I am greatly indebted to Bet Moore for editing my writings.

There must have been 40 of us in the grade 2 class at Park Street, but somehow I learnt to read. The first book that I then read must have been one of Grandpa's that I found in my bedroom (he had died a couple of years earlier). It described how they had built some of the most notable bridges of the world — stringing the cables back and forth for the Golden Gate suspension bridge of San Francisco and building caissons so that they could dig the foundations for Scotland's Firth of Forth Railway Bridge of contrasting cantilever design. Reading it left me with a liking for learning, with a particular passion for physics.

Although only a bush engineer, even in my dotage I like to use physics when fixing poorly swung gates, or when erecting new ones so that sheep cannot get trapped behind them as the mob passes through.

Boarding school

Following the term at Park Street, in February 1939, at the tender age of 10, I was sent as a border to the Geelong College, which despite some high spots was a moderately miserable experience. To understand what follows I had better relate for my readers the circumstances of my birth, which was much more formative than my lack of childhood playmates.

My misery was the undoubted legacy of the Rh factor, of which I had been a victim at birth. It certainly bolsters my self-esteem to think that there was a reason for my having been such a sop.

I was a sickly child having had a battle with my mother's immune system. Her blood group was Rh-negative. I was her second son and therefore vulnerable, she having been sensitised to Rh-positive blood by her first pregnancy. I was a rhesus baby suffering from haemolytic disease at birth as my Rh-positive red blood cells leaking from my placenta in the womb caused my Rh-negative mother to protect her body with preventive antibodies — which had a woeful effect on my body. Rhesus haemolytic disease was only identified medically in 1937 when I was in my ninth year. The name comes from the particular breed of monkeys thought to have a similar abnormality. Fortunately the effects on me were minimal otherwise I may not have been here to tell the story, for the syndrome could have been fatal. I certainly had much sickness in my early childhood, and in my teens I suffered from perpetually running sinuses which earned me an abhorrent nickname. I wonder whether the Rh factor also affected my physical strength and not only that but my psyche too, making me more easily put upon by one older boy when a 10-year-old in my first year at the College. As a teenager did it make me prone to being bullied by others against whom I had no defence?

I was not particularly strong for my size and weight. In late childhood Philip Aitken, although smaller than me, was able to beat me at arm wrestling.

Still on about my perceived lack of strength, a decade later in 1951 when rowing 'Bow' on the Yarra River in Jack Edward's four, stroke side was always pulling bow-side around; I was not as strong as John Varley, engineer friend of Jack's, rowing in 'two' seat. Mind you, my three-man on bow-side, although a big bloke, did not train, saying that he was not fit enough to do so. We did manage to get into the semi-finals at the 1951 VRA regatta having failed at the Henley Regatta. Jack Edwards, while not particularly big in stature, was a tough nugget, having stroked Scotch College to victory in the Public Schools Head of the River regatta. He befriended me in Ormond College and with his dear wife Jane remains my very good friend.

Apart from being a good oarsman John V was pretty smart. In the early days of our Jubilee class yacht *Mañana*, 1951 onwards, we raced it in B division at the Royal Geelong Yacht Club. Coming from Melbourne this particular Saturday John crewed with me, never having been sailing before. We boarded *Mañana* at its mooring with five minutes to spare in getting off the mooring before the 10 minute warning gun as the rules required. With John Varley following my instructions perfectly we did so, then crossed the starting line on time with the burgee flying and the spinnaker set to the following breeze. Smart work. I hope you enjoyed it John: you never came with me again but perhaps I never asked you, as I had completed my university agricultural science degree and left Melbourne. After that digression we come back to me and to the possible effects of the Rh factor during my boarding-school days following those at the state school.

Cripplingly shy I spent the first seven years at the College trying not to be noticed. I would have loved to be part of the rowdy crowd of boys, but shyness kept me alone. In my first year at the College Prep School I would wait until the rowdy ones had relieved themselves at morning recess time before myself going to the toilet, which was an isolated a brick building.

Then a boy older by two years would follow me, force me into a toilet cubicle and bum-fuck me no less: not just a bit of playing around with my privates. His abuse of me seemed to occur about once a week.

I was too humiliated to tell anyone; yet I continued to enjoy learning. Having learnt to read in grade 2 the previous year, I was taken up to grade 5 by marvellous Mr Dunkley. Then he went to the war and came back a wreck. So much for the glories of soldiering. I have now been enabled to move beyond my humiliation and confess to June. On my eightieth birthday I made it public suggesting that if I were to meet the culprit in a toilet I would have to struggle between forgiving him and castrating him. Everybody gasped at my vehemence. Although only having had my body penetrated by the back door, I have emerged triumphant. I can now say to anyone who has been abused that it was not you who was debased by it but your abuser! You may stand tall. You are allowed to be angry.

Then the boy with his precocious pre-puberty sized penis moved up to the senior school and the abuse ceased. Thinking that it was all par for the course I let the memory of it slip from me for 40 or 50 years until I made a conscious effort to forgive the critter and the several bullies of my teenage years. A sensitive boy, I shed tears easily and in my teens several stronger boys took delight in making me cry. One of them was an erstwhile friend. There were two others who, when I walked into the senior day common room, would run up and punch me in the solar plexus to make me cry. It was not a loud sobbing but a quiet blubbing. I do wonder whether allowing myself to be put upon in these ways was partly due to my skirmish with the Rh factor. My inability to throw a ball or a punch might just be due to a lack of quick twitch muscles and not the after-effects of the Rh factor. Nowadays I cry for the poor and oppressed of the world and actively seek to help them. When I shed tears it is because things have unexpectedly gone right for them.

Although I had done an officer training course for the Cadet Corps during the school holidays, in my ninth and final year at the College I was not of sufficient leadership material to be given a one-pip commission as cadet lieutenant; so they made me the Sergeant Major of B company, and from being shy I quite enjoyed telling the troops, mainly younger boys, what to do.

The leader of the Cadet Corps band asked me to instil some discipline into its members by way of marching sprightly and in time.

I remember trying to teach another boy how to march, swinging his arms in the opposite direction to that of his legs. He was trying to march by swinging his left arm and left leg forward together. Much later he accused me of being a bully: perhaps he was right. I do not seek leadership positions but when they come my way I enjoy telling people what to do, if that is bullying. Let this part of my story be a warning to all. It is great to have ideas but we need to be sensitive about how we present them to others. I have not always been so.

Water and windmills and rowing

Another digression. During the 1936 year of riding our ponies to school with John, Dad took us out of school and up to where they were drilling a new bore adjacent to the Mount Mercer Road on the rise near the cattle yards. From there we could see the beautiful blue of flax in flower in the '32 Acre', one of Dad's initiatives that didn't work. (The paddock is now part of East McCrimmons.) The need of a bore for stock-water says something about the climate in the mid to late 1930s with there being not enough run-off to fill even our pothole dams. They were all we had until a quarter of a century later Max Currie 'walked' two of the largest crawler tractors across country each pulling a big scoop. Every farmer he approached he persuaded to build new dams. He enlarged the house dam where I used to dog paddle across the corner. Using the spoil from the excavation he built a bank to hold water four feet above the 'table', the normal ground level at the bottom of the slope. When full I estimated it would then hold 4 acre-feet or 1 million litres of water, a 30-fold increase in capacity.

I seem to have spent quite a bit of my adult life climbing to fix windmills, including the one on the 1936 bore. It and another in the Old Cultivation only accessed a perched water table about 20 m down. With the rainfall becoming less as the climate changes, even less than what prompted Dad to sink the bore near the cattle yards, they have now both gone dry. When there was water to be pumped it was necessary to service them frequently, seemingly every year. They were lift pumps, so the pump itself had to be within cooee of the water table and the foot valve actually in it. To fit new leather washers to the pump and foot valve it was necessary to pull up, one by one, the several lengths of pipe and pump rod to get access to them. I used a simple block and tackle to withdraw each length a few feet at a time and a heavy clamp to hold the pipe between each pull.

As the shelter belts planted by Dad grew around *Glenspean*, the windmill at the house which lifted rain water from the underground tank to the overhead tank became sheltered. The household water supply had to be augmented, first by a hand pump, then by a petrol-motor driven centrifugal pump. The windmill became superfluous. I extended its legs by 10 feet and took it to the house dam to pump dam water to the overhead dam-water tank on the higher ground north of the sheep yards. Many a time did I spend 40 feet up above the house dam to service its gears and the differential head which enabled it to pump water on the 'down' as well as the up-stroke; also to fix loose blades on the 10 foot diameter wheel.

All that was in the future to my boarding-school days. If my perceived youthful lack of strength was due to the Rh factor I seemed to have just about conquered it in adulthood, for as I tell it I did some fairly amazing things on the farm, like repairing windmills, or backpacking a motorised mister to spray out variegated thistles on the steepest part of the creek bank, a 25° slope.

Rowing was one of the things that made life bearable in my final years. I rowed in a tub four one year, then in the fourth eight: the College only boated four eights. In 1947, I held down the 'seven' seat in the Geelong College second eight and really enjoyed training with the crew; not so the actual racing. Our coach had the bright idea of starting us in the full forward position. At the start of the heat I gave an almighty heave and came off my seat and had some very uncomfortable seconds trying to get back onto it and at the same time keep time with the stroke. We lost the heat and there were no repechages. That was the end of me being part of a schoolboy crew on the Barwon River; but I quite often went out in a scull to row by myself. At Ormond College I enjoyed rowing in the second eight, but we deemed ourselves to be 'the gentlemen' and did not take it very seriously. Rowing with Jack Edwards in a narrow racing four I found the balance very difficult after crewing in wider training boats. Otherwise it was enjoyable.

The House of Guilds at Geelong College was another place of peace. In my first year my shorts were falling down, so I plaited a six-strand kangaroo-hide belt. Over the years I did some nice sewing, such as a leather camera case, but unfortunately a bridle I made was big enough for a draft horse.

I loved George Logie Smith's Glee Club. 'Boof' was a robust young man, not at all the type that you would associate with being a music master. He played football with the Geelong Amateurs and as a two-pip lieutenant he ran the Cadet Corps, nominally commanded by his friend, JH Campbell, a genuine captain from the First World War. For over 20 years Boof Smith produced Gilbert and Sullivan (G&S) operettas. We practised the songs weekly and towards performance time in third term he claimed whole days for practice. When in the staff common room other teachers bemoaned the loss of their teaching time it was JH Campbell who stood by his friend. Performing in a G&S was quite transforming despite the fact that I never learned to read music, and can even now sing neither in tune nor in time. Yet 70 years later I can still sing some of the principles' songs. It is a great antidote when feeling sleepy at the steering wheel: my singing would keep anyone awake. I was singing as a little girl in two G&S shows before my voice broke — *HMS Pinafore* and *Pirates of Penzance*. After several years I returned as a first bass in *The Mikado* and

The Yeoman of the Guard with Fred Elliott assigned to stand behind me singing in my ear; but I never listened.

College staff may have gone out of their way to ease the life of this exceptionally sensitive boy. Tolerating him in the Glee Club was possibly an example. At the prep I had been asked to drum the boys as they marched to morning assemblies in the Morrison Hall. A senior boy David Woolley showed me how to get two beats out of one stroke of each drumstick, by bouncing it. Not so the College Principal. If in the Glee Club I did not listen to the music, neither did I follow the flight of a ball. I never watched it. In his last days as Principal, *What-Say* Rolland was walking past and called me over as I was playing kick-to-kick in the goal square.

'I say, couldn't you see that the ball was going right over your head?'
'No sir.'

End of conversation. Frank Rolland had been a pretty good tennis player in his time and knew how to follow the flight of a ball, but he made no attempt to teach me.

I taught myself to follow flight when hitting a ball against a wall in later years. It was like that with gymnastics. It was only as leader of the Elaine-Meredith Youth Group that I learned to turn a somersault — laboriously, by walking up to my hands.

At Geelong College in 1946 I matriculated in English Expression, pure and applied maths, physics and chemistry, with a second class honour in one of them. Brian 'Podguts' Lester was a magnificent maths teacher. It was in his classroom that I became friendly with Neil Everist, although he was only a 'greasy' (a day boy). June and I maintained friendship with him and his beautiful wife Jill until he died in his sleep, having the previous day been fit enough to do laps of the pool. He was so much my superior in all respects — he and his older brother Ian were international oarsmen and renowned architects. As College architect Neil was responsible for much of its building, including the design of the Mossgiel boarding house, completion of the quadrangle, building the sports centre and the enlargement of the Norman Morrison Hall.

I was deemed to be still immature and was sent back to College for another year during which I did two histories with JH Campbell, geography and French with Bertie Keith, and English literature. My initials are NG. Fido Tait the English teacher came marching down the aisle between the desks brandishing my essay labelled N <u>B</u> G Cameron. He never did teach me how to write an essay; nor what I should have written about Prospero, the hero of Shakespeare's *The Tempest*. Much of my geography and of my grammatical skills come from the teaching of *Bertie* in the Sub-intermediate, year nine.

In all my nine years at the College I had just two notable impulses. In the Intermediate, year ten, 'angry boy' was attacking 'fat boy' with flailing fists. Something said to me, 'Get in between them,' but I did not. I don't know why. It was not because of fear nor of the fact that I am unable to throw a punch myself. (The ability to throw a punch and to follow the flight of a ball are two things necessary for joining in the fun and games at school. Lack of them made me more of a loner.)

In my final year at the College, for the winter mid-term exeat, 1947, a dozen of us piled into a furniture van with our skis and went skiing on Victoria's Mount Buller, me with my homemade skis.

Yes! Campbell Curtis and I both made our own skis. He took them down to Higgs the timber merchants to get the points steamed and curved upwards. I bought steel edging and quarter-inch screws from Jarmans sports shop and shaped the toe pieces from some heavy galvanised sheet metal which I found. I then spent ages ironing Stockholm tar into the soles of the skis but I need not have bothered because the tar soon came off, which enabled me to walk straight up moderate slopes without herringboning. It was a heavy snow year, typical of the late 1940s, and on the approach road to Mount Buller the snow was down to the White Bridge where it crosses the Boggy Creek which rises near the summit. Having abandoned the furniture van, which was our transport, I was trying to catch up with the fastest skier group, leaving another group of slow ones behind. It was dusk when I arrived at the 'summit' cattlemen's hut, but instead of going inside to claim a sleeping space I stayed outside waving my torch around to guide in the slow ones. Whether it made a difference to them I do not know because nobody ever mentioned it. I had failed to protect the fat boy from flailing fists; on this occasion my action may have protected the slow skiers from spending a very cold night in the snow.

✱✱✱

Eleven years later, having amongst other things been emancipated from the slavery of self and freed from the fear which had held me bound at school, I was in a position to help save a novice skier from freezing to death. Leaving our little girl Jane in good care, June and I had spent a week of dismal weather on Mount Hotham when a continual blizzard confined skiing to the upper part of a ski run where it was protected by snow gums. Our host at the Drift Chalet was Lindsay Salmon, who was also the ski instructor. He had spent his wartime years teaching soldiers to ski on the heights of Lebanon.

On our last day at Hotham, Lindsay was going to meet an inexperienced party being brought in by Fred Ward on the approach coming up from Omeo in Gippsland. Now a thriving village, Dinner Plain in 1958 was just a flat area amongst the snow gums. It was as far as Fred could get his passengers in his four-wheel-drives; but was still 8 miles down from the Hotham Heights Hotel and the nearby Drift Chalet. Lindsay accepted my offer. I was tardy getting away but hurrying to meet him I passed on my way a lady collapsed in a heap of exhaustion. It was dusk by the time I had rendezvoused with Lindsay who, with his headlight lighting the way, was leading his guests in file. He asked me to bring up the rear, shining my torch forward to light the road. When we came to this exhausted lady with her man, Lindsay called me forward. I left my torch with a young man to bring up the rear.

Lindsay and I lifted the lady to stand upright on her skis between us, each with an arm around her and with one of her arms over each of our shoulders. We all three walked together for a mile or so up the slowly ascending road until we came to the Slaty Cutting about a mile short of the ski village. There the road follows a narrow ridgeline where the southerly gale had carved sastrugi to block our path. Although the ridges would have been nowhere near a metre in height they necessitated the three of us side stepping up and down each one on our skis. Having negotiated the sastrugi, it was another kilometre before taking the downhill slope to the Drift Chalet. From its door the girls of Lindsay's staff were anxiously scanning the track. With just about my last gasp I said to them, 'Grab her!' They took the lady inside, but I sat exhausted in the drying room for at least half an hour before I had strength to go in.

In the morning the lady said to me, 'You saved my life,' when in fact it was mainly Lindsay's doing. I just happened to be in the right place at the right time to help. Thinking about it 60 years later, I now see it as a privilege to be able to help people, no thanks being required.

Discussing June's and my relationship with my mother I have already related in an earlier chapter how my prompt action saved June, myself and two friends from drowning when *Mañana*, hit by a violent wind change, would have capsized and sunk without trace. Even now I shudder to think about it.

Not so long afterwards I had taken Bill and Elwin Hermiston from near Deniliquin for a sail. Having moored on return we were standing on the deck of *Mañana* as I untied the dinghy to take us ashore. I saw two

10- to 12-year-old boys set out from the shore in a dinghy using a piece of matting as a sail. Next their little boat had capsized, probably because they had their weight at the 'sharp end' as they held up the mat to the breeze. They were floundering in the water. I grabbed the horseshoe flotation device which it is obligatory to carry on board, and in next to no time had my dinghy skimming over the water towards the boys. Coming to the first boy I handed him the horseshoe saying, 'Hold that!' Then I went to the second boy, swinging my dinghy around so that he could climb in over the broad stern. Then I went back and picked up the first boy and took them both to shore where they ran away. I returned to *Mañana* and took Bill and Elwin ashore. For a long while I kept a ducky little straw hat worn by one of the boys as the only memento of the event.

Apart from displaying a certain amount of boatmanship I had done nothing heroic. I had learned to respond to situations very quickly when John and I sailed *Royal* in the earliest days of the Sorrento Sailing Club. *Royal* was an old-fashioned clinker-built 12 foot cadet dinghy with bow sprit, dipping lug sail and dagger centreboard. In cross-section it was shaped like a saucer. It had no half-deck so that if the lee rail went under it scooped up water and you were done for. We only bottled four times. On the first occasion, having been taught the rudiments of sailing by Bob Buntine, I was at the helm. Thereafter John became the skipper. We learnt to be very quick in response to changing wind and to the challenges presented by other boats when racing.

Bob helped me on my journey by asking me to sail in *Royal* with him and tall Don McMillan, school captain. On opening day at the Royal Geelong Yacht Club, we sailed out in front of the crowd only to be hit by the Yarra Street puff as we went about. Big Don's feet slid from under him and we bottled. Dad bought *Royal* from the Club and had it taken to Sorrento.

Bob's action in giving me that invitation has been like a stone dropped into water, sending ripples in ever-expanding circles. As a result I introduced Dad to sailing, brother John, his and my children, and many others. So it goes on except that the ripples are getting rather faint when it gets down to our grandchildren. Some prefer surfing. The ripples carried those two drowning boys to shore.

I just wish that in our early days as school-leavers I had had the attitude and aptitude to help EG (Egg) Roberts. I had neither when he put out his hand for help. He had invited me to join him on his farm near Yendon, but I did not follow that up by asking him to make a return visit. If I had given more of my time to him and had had the listening skills, he might not have committed suicide. A beautiful painting he gave us hangs on the wall as a memorial.

Briefly taking the story back to 1947: in that year **John MacDougall Stewart and I met up again when we were allowed to do** our evening study in a separate room from the other boarders. He had the bed alongside mine in 1939. It was in the rafters of his garage that I parked my homemade skis after some mutual escapade, possibly with the University Student Ski Club. He was a resident at Ormond College, too. I wish I had those skis now as a memento of my ability to make them a matching pair.

6 University life

University life gave me a new freedom. The teaching staff of the agricultural faculty boosted my ego by addressing me as Mr Cameron. Max Butler and I were put together alphabetically in the practical laboratories of first-year chemistry and physics and we remain lifelong friends, although our interests have diverged.

We spent our second, practical year, at Dookie Agricultural College where I learnt the important skill of driving an eight-horse team. The two girls in our year were housed with a staff member. We 14 men became close: but not without tricks. Six of us were housed in cubicles partitioned off from the old gymnasium. The others decided that we should have a flower garden outside, the position of gardener being determined by drawing straws. I drew the short straw. It was only years later that Bill Parsons confided to me that all the straws had been short.

Each fortnight we were assigned to different activities at the College, days on the farm alternating with days of lectures. It was the turn of Max Butler and myself to bring in the cows in the morning for milking. Pat Arundel set his electric alarm clock to get us up. We duly arose and brought in the cows only to look at the clock in the dairy and find that it was one hour too early. We tried to take the cows back but they refused. The concrete cow yard became filthy. When we arrived home, Pat's clock was showing the correct time but funnily enough he seemed to know that we had gone out one hour too early.

As freshmen (scum) at Ormond College we went through an initiation when we had to learn the particulars of the 149 other students — their name, study number and faculty. The returning students, not scum, were designated by a title. The medical students were 'Dr XXX', the engineering students 'Mr YYY', the theological students 'Rev ZZZ'. Once initiation was over we became a fraternity.

Which reminds me that five days a week I had four hours of lectures in the morning followed by three hours of lectures or prac work after lunch. A lecture after lunch would find me ZZZing. I would try writing with my eyes shut until my right hand slid off the page of my notebook.

I did become very adept at listening to a lecture, at the same time abbreviating it into notes. I never missed anything and those notes became the basis of swotting for exams, there being as yet few up-to-date textbooks in those postwar years. In the third year I took out the Prize in Agricultural Chemistry, and in fourth year the Exhibition in Agricultural Engineering and Surveying but I failed to invest the ten-pound prize-monies in any lasting memento like a microscope.

My university days completed by December 1951, lacking 'umph' I returned home as expected and immediately became involved in the workings of the farm, including yet again the hay harvest. By the following February, with brother John marrying Beb then honeymooning for a month in New Zealand, and with Mum and Dad embarking upon their world trip, I found myself in the driver's seat at my farm where I have remained until gradually relinquishing the post to Sandy and his wife Julie.

Up until this point I had not shown much initiative, but it was my suggestion to the Dean, Professor Sam Wadham, that I should do something with my degree that contributed to a great change in my life trajectory, through my involvement with cattle in Prof Terry Robinson's world-famous Glenspean Experiment.

A much greater initiative on my part led to the most momentous and continuing event of my life. It was the winning of June to be my wife. On the third time of asking, at about two-monthly intervals, and with her poised looking down from the cliff tops of the Sorrento southern shore, she 'Couldn't see why not'. She has been seeing why not ever since.

Concurrent to both of these there was something else momentous which made me a not entirely unsuitable husband to her and father to our family. It was my miraculous spiritual emancipation from so many of the inhibitions which had held me captive, the subject of another book which I may call The Twisted Tree. It was not my initiative at all, but that of the Someone or Something who or which had kept an eye on me after, at the age of ten, I had turned my back on the man Jesus. I had been sent to Sunday school just once. Inside the cover of the book that they gave to me was the invitation, 'When you give your life to Jesus sign here.' I had never been to church and had only the faintest idea as to who Jesus was or is; but I signed and then hid the book which was never seen again. I had my back turned to Him for 18 long years, even when associating with Ormond College theological students. I had told myself, correctly, that nobody believed in Him and that anyway He was a wimp — wrong. Mark's Gospel reveals Jesus as a practical and compassionate man, clothed with a power greater than Himself. With great courage He ventured into Jerusalem offering peace and truth to the capital city. He did not shrink from the pain and humiliation of crucifixion when the religious and secular authorities colluded to murder Him. Hardly a wimp.

I had turned my back on 'Him'. Apparently 'He' (present in spirit I suppose) had not turned His back on me. Thereafter, 18 years of preparatory experiences had me ready to receive His proffered friendship. More or less concurrent with the Glenspean Experiment and my marriage to June, He called me into the fellowship of His church around which I had hovered; but first I had to respond by offering back to Him what I valued most — money! As I signed a weekly commitment to the church far beyond my inclination, all the little strings which had held 'Gulliver' down were swept away. It was an instantaneous release as I signed; yet it is an ongoing miracle.

The combination of these momentous things in my life has led me to spend more time on family and community activities than on running the farm.

June, August 2018

7 Agricultural scientist Neil: returning to the practicalities of farming

When in 1950 I got home for the summer, fresh from my second, 'practical', university year at Dookie, the paddocks that had been closed up since the beginning of September had already been cut and raked. The team of us would go out to bring in harvest. At lunchtime we would gather around the wooden lunchbox – which was by no means airtight or dustproof. It contained the basic ingredients, bread, tins of WHAM, jam and so on. As the youngest it was my job to get some sticks, make a fire and boil the billy. While it was bubbling away I would throw in a small handful of tea, then tap the billy on the side, creating little wavelets to help the tea leaves to sink. The billy was large so it was not filled to the brim, meaning that when tilted to one side it was possible to grasp its other cooler side rather than the handle and so pour the tea into the pannikins.

Cattle and haymaking: the Roto-Baler

If you want cows to make milk for their calves you have to feed them.

As I was finishing my 'practical' year, Dad bought an Allis-Chalmers Roto-Baler from the chappie Mitchell who sold them, and was also a friend. It was pulled by 'big Allis', a 40-horsepower tractor, which had a foot clutch for operating the power take-off and a hand clutch controlling forward locomotion. With each of my birthdays falling on 4 December, many a one have I spent dealing with the idiosyncrasies of our Roto-Baler: just 300 bales was an average daily tally.

Over the years, with light showers of rain coming once a week in late spring, my brother John developed a system of raking to avoid damp patches of the windrowed hay getting caught on the points of the Roto-Baler pickup mechanism, or from wrapping around its feed-in rollers. His system involved raking the hay into twin rows 1.5 metres apart; then just before baling raking them together so hay, which had been damp from lying on the ground, was exposed to the sun and air. We used Vicon rakes, each with five finger wheels. John put two of the finger wheel rakes together, one turned upside down as it were, to make a V-shaped twin-rower rake. It involved dismantling the finger wheels of the second rake and re-assembling them so that tynes were dragging along the ground, not digging into it. The V-rake rate had additional rake wheels in between the 'V', raking outwards to make a clear path for the final row. His configuration is one which has now been almost universally adopted by commercial manufacturers. However, his system had a feature not adopted by them: his V-rake left twin rows to be pulled together by his Super-rake with four spring-loaded rake wheels, when the drying was complete. As an innovative thinker of practical solutions to problems he follows in our Dad's footsteps.

Quadroll prototype built by John, c.1973, to provide the means of moving large round hay bales; still used daily at Meredith Dairy, now carrying shorter bales

A peep backwards in time

I have made our haymaking all sound like hard work; however, it was a lot less so than previously, when Dad had the Barn paddock raked into windrows with the trip rake, which was then driven along the rows making little heaps or 'cocks'. Dad then drove along the rows with two horses and the push-rake gathering the cocks into heaps from which the hay had to be pitchforked into a stationary baler. Having made heaps in the Barn paddock he went to push up more heaps in the '55 Acre' – but at the sight of so much more work ahead Pinky, one of his two horses, dropped dead.

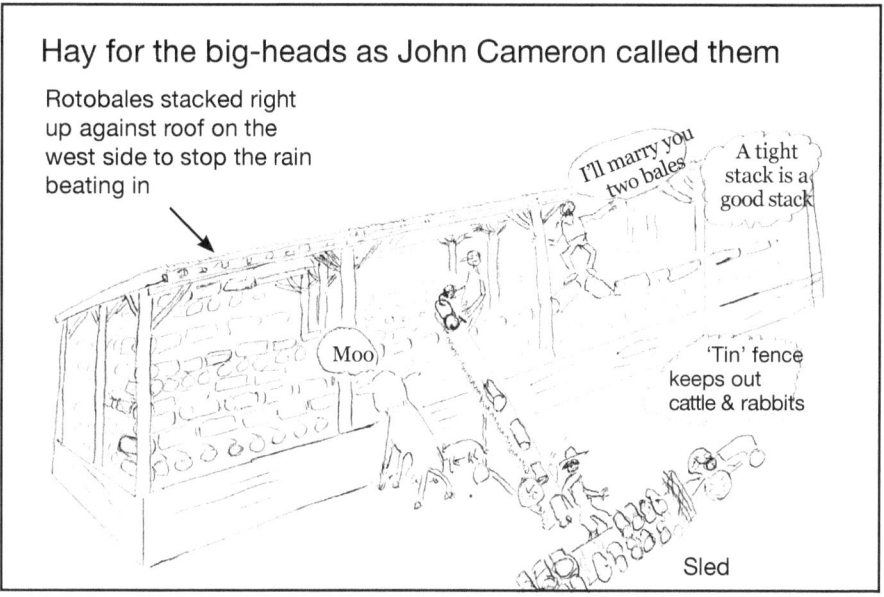

The small square bales were then loaded onto a rubber-tyred spring cart pulled by Blossom and taken to a skillion attached to the barn. Dad had built it to be convenient for his 'sub and super' sowing-down program. As it was located on the far side of the creek from home, the carriers bringing in the necessary seed and superphosphate did not have

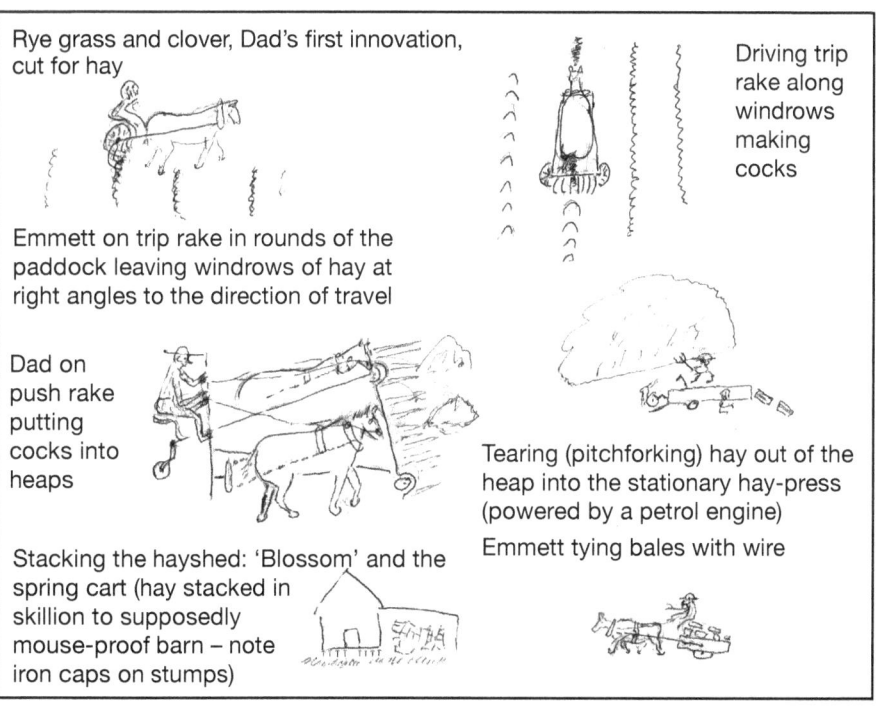

to negotiate its steep sides. Later the barn was used for the bulk-storage of oats. Dad had built it to truck height on wooden stumps with old roofing iron used as caps, supposedly making it proof against mice.

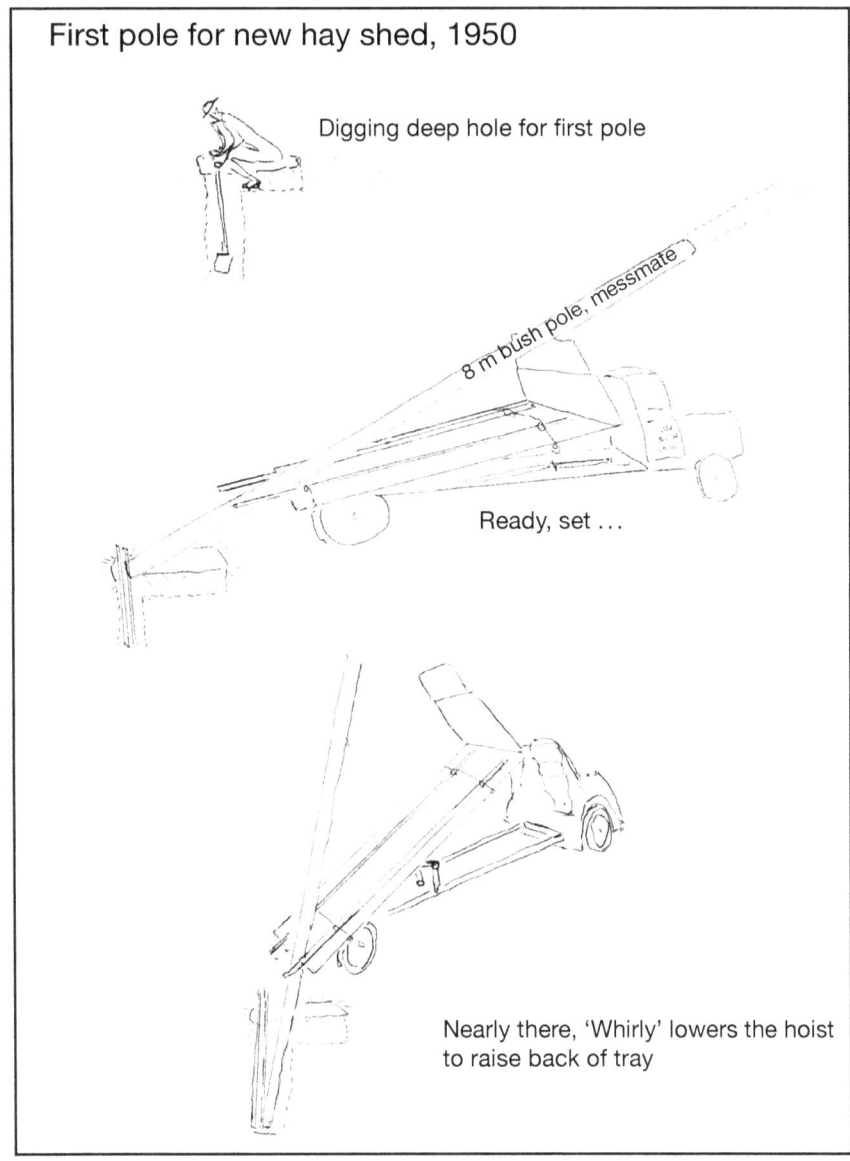

First pole for new hay shed, 1950

Digging deep hole for first pole

8 m bush pole, messmate

Ready, set …

Nearly there, 'Whirly' lowers the hoist to raise back of tray

We completed building the 18 feet high hay shed in the south-east corner of the 77-acre adjacent to the barn. It was supported by bush poles sunk some five feet into the stony ground. It was necessary to dig a 'window' in which one stood in order to get the necessary depth with crowbar and post-hole shovel. Then 'Whirly' Bill Wiseman would bring his truck to the window with a pole resting on the hay frame at the front and slanting down over the rear of the tray towards the hole with its access window. We would slide the pole down his truck until the butt end was in the window, resting against the back of the hole; then Whirly would reverse to bring the pole into the vertical position. He had twin planks roped longitudinally to the back of his truck acting as guides and two more planks at the back of the hole to encourage the foot of the pole not to dig into the side of the hole but slide downwards. With 15 poles in place for its four double bays, we filled the embryo shed to a suitable height from which we could reach to build the roof. The shed had a skillion-type roof rising by 3 feet from west to east. Additional posts between the outside rows of poles allowed for a 5-foot high wall as protection against vermin (including those black 'big-heads' suckling their automated self-propelled milking machines).

Having no longitudinal strings to grasp, as square bales do, we used a large bale-hook in the right hand, sometimes two of them in both. The cylindrical Roto-Bales were 3 feet long and about 1.5 feet in diameter. Thus when filled to the roof there were 12 layers. To stack the bales so high it was necessary to build using a repetitive 4-bale-high pattern in which every second layer had a bale placed longitudinally along one edge of the stack or the other. At the corners they were prevented from falling out by being wedged against the pole.

We built a sledge 5 feet wide which was towed between the rows of bales with the man each side picking up the bales and placing them crosswise on the sledge. When loaded to the height of two or three bales it was drawn up close to the long elevator carrying the bales into the shed.

Later on we replaced the sledge with our own truck and a paddock elevator; but for that we had to go around the paddock turning the bales by 90 degrees to be in line with the direction of travel. At the same time we would tie the loose end of string binding the bale.

The beauty of the Roto-Bale system was that once rolled up the hay was proof against rain; nevertheless if the bales were left lying on wet ground they had to be rolled over for the bottoms to dry before carting. It is a system we used for 35 years. We built a second 6-bay shed to the east of the *Glenspean* woolshed and sheepyards, plus a third, 8-bay one in Randall's. There we selected a site with deep clay subsoil and were able to use a rotary digger with extension for the 5-foot deep holes. (During its construction, when doing the property valuation for Shire rating purposes, the valuer commented, 'It is only a temporary shed', i.e of no value for rating purposes. Sixty years later it is still in use for storing big square bales – food for goats.)

As school-leavers in 1977, son Simon with his cousin Andy filled the Randall's shed with Roto-Bales. It was the finale to the Roto-Baling era. That hay remained unused for several years but came in handy during the 1982–83 drought.

Half a generation earlier than that I had taken to leaving some rolls on the ground to be picked up directly for feeding to the stock over the early autumn period. With Jane aged three steering, and the truck in low ratio, I walked beside it as it went along the rows of bales and threw them up two-high onto the back, for feeding out the same day. It avoided the expensive chore of carting-in. If Jane's steering was not too good I could lean in the open window and grasp the steering wheel myself.

Big rolls are now all the go, but as I wrote this in November 2017 a relative told me that he was making small square bales for sale, presumably for people with hens or horses. It is best to cart them the day they are made because they are vulnerable to stormy weather.

✳✳✳

My pony Creamy had never kicked me, but she had a reputation for kicking out of the loose-box any unfamiliar man sent in to catch her. (Cousins 'Puss' and 'Plugger' Grant were two who received a double whammy from her, but worse was to come for them. They are remembered as Meredith's war heroes whose lives were sacrificed on our behalf during the Second World War.)

In the early days, square bales were tied with wire. This became strewn around the paddocks during feeding out. Creamy was allergic to wire, so when riding wherever I needed to go, I used a stick to scoop up pieces of wire as I went. On one occasion in the early days of the war, with Dad in the forces again risking his life for us, Mum had bought a wagon-load of oaten hay from our neighbour Lex McNaughton. John and I were deputed to ride out to Billings Point and help unload it. I tied Creamy to the remainder of an old fence, just barbed wire strung between the posts, but she pulled back and with a length of wire trailing behind made off until she stopped, petrified. Her reins were still tied to the barbed wire and her tail was tangled in it. As a young teenager I somehow or other managed to free her without being kicked.

In addition to grass hay we also made the oaten variety, a labour that had its ups and downs. The oat crop grew up tall. Before maturity it was cut down by the reaper and binder then bound up into sheaves which fell to the ground. The men stood them up, two leaning together, then two more and finally about eight in a stook. When the drying was complete the stooks were knocked down and the sheaves pitchforked up onto the wagon or truck from where they were thrown down to the stack builders who made sure that the haystack was well hearted up so that the straws sloped down and outwards to shed any water which penetrated stack. Stack building was another of old Emmett's skills and brother John learnt it from him.

Stones

To protect the stacks from soil moisture they were frequently built onto a bed of stones sledged in from the paddock where they had been loosened by the ploughing before the crop was sown. Because of the stone a 'stump-jump' disc plough was used. It had been invented to cope with Mallee stumps, the lingo-tubers from which trees of the Mallee group of Eucalypts send up multiple trunks. Mallee roots were sold in Melbourne as premium firewood. Travelling through Melbourne by train one would see wayside wood yards with their stocks of Mallee roots.

In my time as joint manager with John, many paddocks were needed for mowing pasture hay to feed the sizable herd of cows and calves. To protect the mower, paddocks had to be free of stones. Leading a team of three or four men, many a stone did I pull out of the ground with a bale-hook. Larger stones we prised out with a crowbar and if that failed we used a sledgehammer to smooth the top so that the mower could pass over without damage. One poor lass doing her work experience from school spent one entire week helping me pick up small stones from the '55 Acre'.

Sheepman Neil: early years mustering and later

I do not remember my first mustering job but it must have been after Dad's return from the Middle East at the end of 1943 that he would arrange shearing, crutching or other sheep work to coincide with my school holidays. I brought in a mob of sheep across the creek after breakfast, another after morning tea, and two more after lunch. The creek paddock extended right across the access route: south-north from the sheep yards to the barn gate, and east-west right across the farm.

I would ride out towards the horse crossing with dogs, including the rabbit pack, and old Bill lagging behind. When Bill got to the crest of the hill he would stop to relieve himself and once we were out of sight he would turn around and go home.

We always dismounted and walked up the steep bank on the far side of the horse crossing out of care for the horses and because it was more comfortable. Bringing a mob of sheep home down that steep bank I would sit on the crest and watch the mob as it crossed the creek. If there was another mob of sheep in the Creek paddock it had to be chased well out of the way, easy enough when you are riding. In that era I never did get two mobs boxed, i.e. mixed; however, I did in later times when I was manager of that end of the farm, when the separation of two mobs was quite extraordinary. I remember taking a mob of crossbred ewes outward bound; but before we had reached somewhere near the Barn gate a mob of Merino wethers came galloping up another section of the hill, around a corner in the fence line and joined my mob. Disaster? Well, no. After a while the two mobs separated themselves, with a little help from me, and I was able to drive the wethers away before letting the ewes through the gate into the Barn paddock.

It was a morning's work to bring a mob from the distant end of the farm, even from across the river. The middle section of the farm, between the creek and the river, was two paddocks wide with sheep in every one; so I would leave old Glenn behind the mob and gallop on ahead to move the resident sheep as far away as possible from my route.

At shearing time John would be in the shed classing the wool clip, whilst I did the outside work including mustering. We took to dipping the sheep against lice off-shears so that none was missed. I would see to that too.

I had various helpers but the most notable was Keith Reher Innes. He was as adept at drenching sheep as anyone I have seen. Having filled the 48-foot-long drenching race he would send his dog Reher wide, away from the race to the front, when it then came along it, barking so that the sheep ran forwards past it and snuggled up to each other so that Keith could fit some more in. Keith then worked from the back of the race along the right-hand side. He would reach forward with his left hand as though to grasp the sheep's muzzle but before it had time to struggle he would slip the nozzle of the drenching gun into the side of its mouth, in the gap between the front incisors and the back molars, with a circular motion up under its upper lip, then pressing down on its tongue so as not to injure the back of its throat. (We had once lost a lot of sheep from infection related to such throat injuries.)

Keith's wife Margaret helped June in the house and would mind the children when they were young. Keith often took Sandy fishing. They worked with us for 20 years, and moved into the brick house we, the Partners, had built for them for even longer than that after Keith injured his back.

Rabbiting

In Dad's day he had excluded the rabbits from two paddocks with wire-netting fences – the North River paddock on this side of the river and the Mine paddock on the far side. It became my youthful job to patrol the fences, using stones to block any run-throughs under the netting. In one place, peering in among the leaves of a bough hanging over the fence, I thought I saw a snake. I didn't wait to examine it any more closely, but jumped a mile and left it far behind.

Some netting of high-quality was once made at Pentridge prison, and post-war, as this and netting from other sources became available we proceeded to rabbit-net the whole farm. Rather than just clipping the netting to the existing fences we laboriously pulled out some of the wires and re-threaded them through the netting to support it. The gates as well had to have netting applied and be set close to the ground so that rabbits could not get underneath. Wherever we rode it was with a rabbit pack in addition to the sheep dogs. Plantations with a cover of grass under the trees was a favourite with rabbits. Dusty our black cocker spaniel made a specialty of sniffing out rabbits squatting in the plantations. One crunch and they were dead.

When myxomatosis was first released, Dad spread it in our district, much to the ire of those who made a living from selling rabbits or rabbit skins. In our district it did no more than halve the rabbit numbers.

John had a horse named after a famous racehorse, Ajax. 'Jacko' was so calm that John could shoot rabbits from his back. On one occasion rabbits that had got through the netting were sitting out after heavy rain. On foot, John and Dad between them shot 120 with their 0.22 gauge rifles. In my day I could walk down to the creek towards dusk and shoot sufficient to feed the dogs.

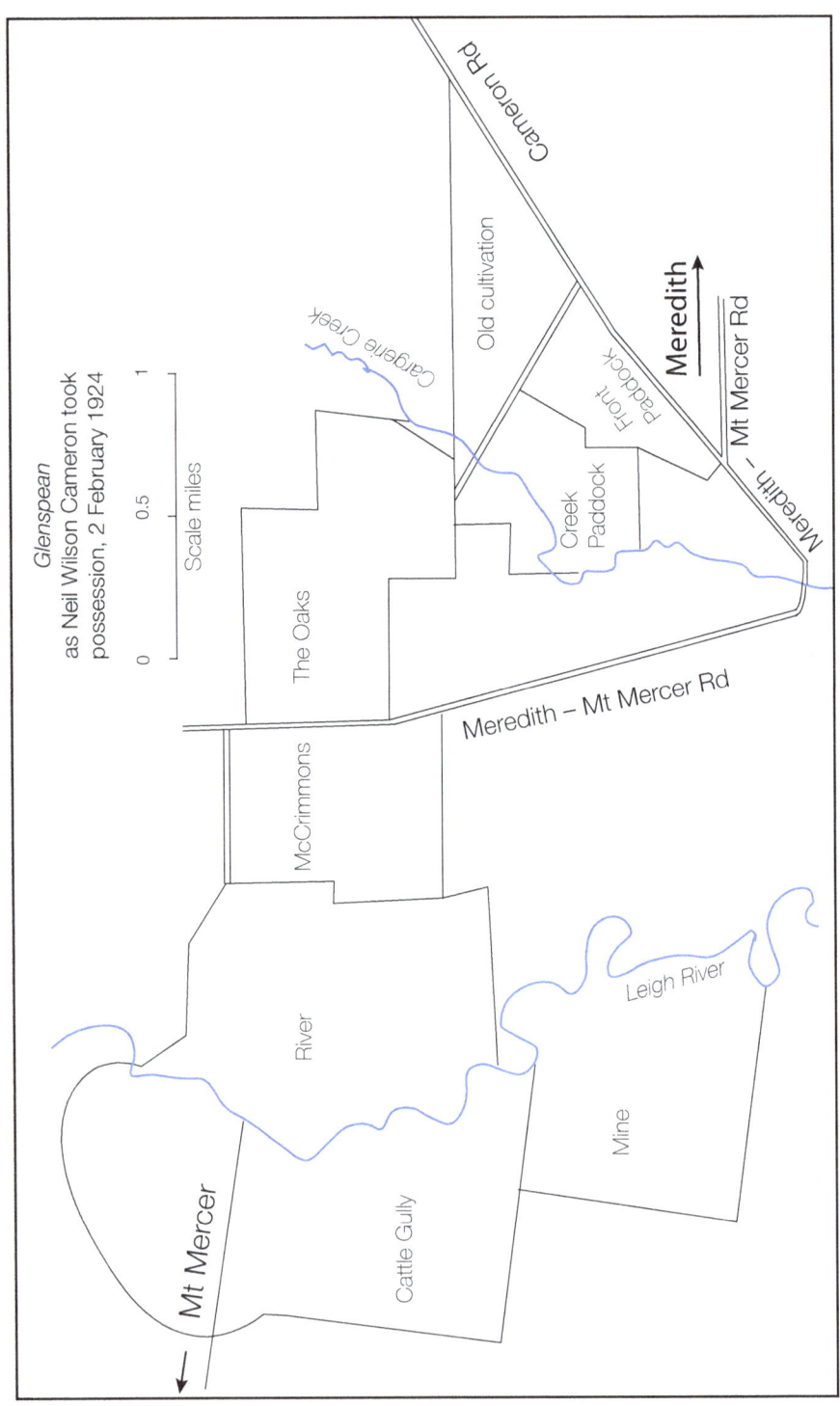

7 Agricultural scientist Neil: returning to the practicalities of farming

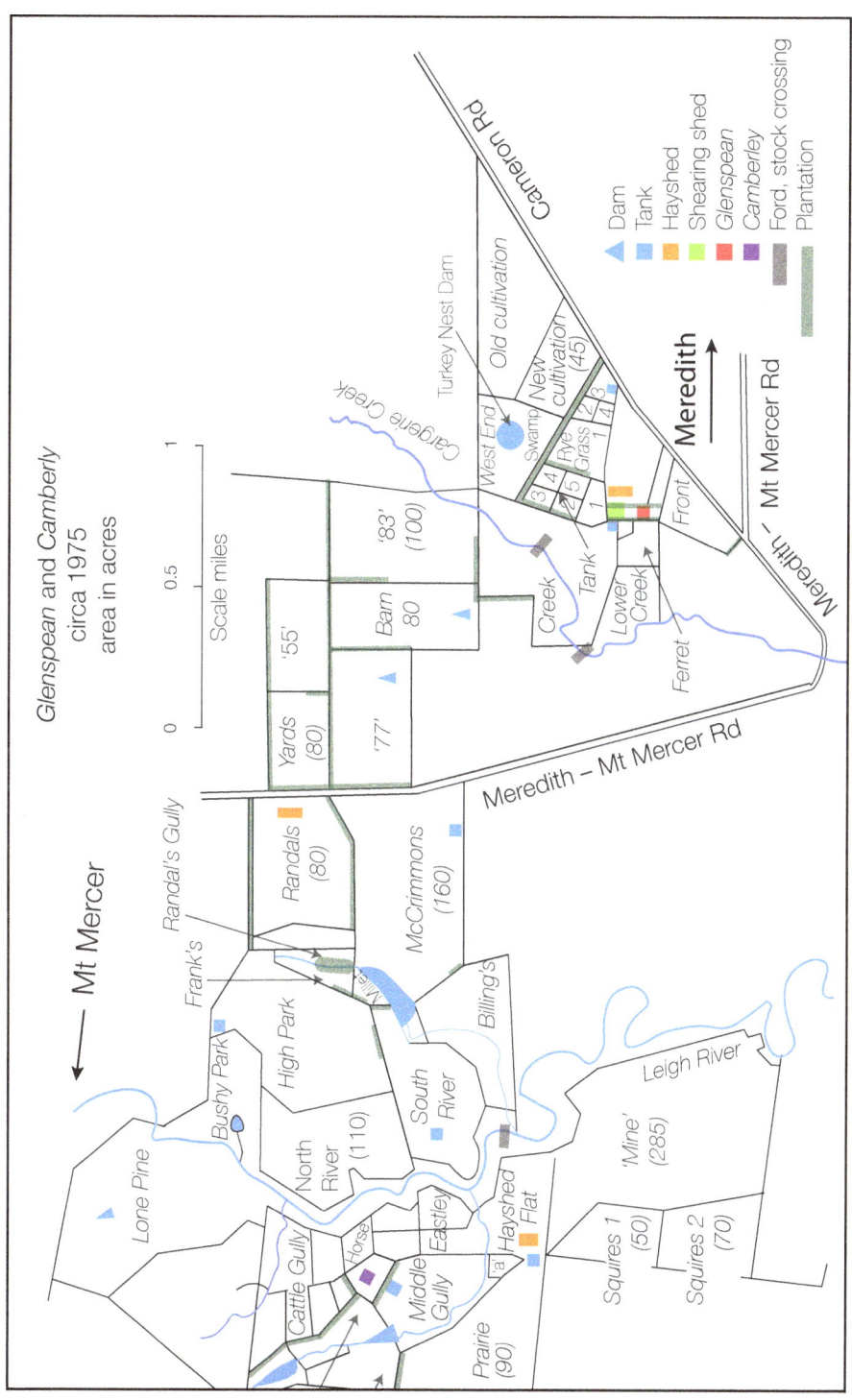

Fumo-flakes (calcium cyanide) were spooned into rabbit burrows, the spoon riveted to a longer wooden handle. It was not too dangerous as the Fumo-flakes would only release prussic acid on contact with moisture in the burrows. My sister-in-law Beb's introduction to 'The Boss', Dad, consisted of his handing her a tin of Fumo-flake and a shovel saying, 'I'll go this way: you go that way'.

Then the war gas, chloropicrin, became available. It is an eye and lung irritant. Mixed with oil it was squirted in and the burrows sealed. Later when it was sold as Larvacide I spent months rabbiting with a team. With my left hand holding a stone or hard clod with which to block the burrow in a creek bank, thus making it hard for the gassed rabbits to scratch their way out, I would take aim, close one eye, half-close the other, hold my breath and squirt with an oil can, backing off as I placed the stone. We had spent a month netting out the river on its Camberley side and then while John was on his honeymoon, and Mum and Dad were heading overseas, three of us using Larvacide spent a month clearing out the rabbits from the creek and hill of what was to become John and Beb's 'Camberley'.

We never had much success poisoning rabbits with 1080 and carrots laid in a trail after several free feeds, but aerial poisoning of the river outside the netting was a great success. It was instituted by the local Vermin and Noxious Weeds Advisory Committee of which I was a member. As a follow-up, with the breeze behind me I would walk along atop the east bank of the river, throwing rabbit bait out and down the hill as far as I could.

Uncle Malcolm, farming at Ruffy in the hills above Euroa, insisted that you must remove the rabbits' harbour. Ripping of burrows and burning of old stumps is the best long-term solution. We did a lot of that on Bungle Gully, the 400-acre sandy block that John had bought cheaply whilst land prices were still under war-time regulation and which he later sold to June. One terrible day the family helped by carrying shovel-loads of hot coals from one stump to another, but June and sister Rosemary had forgotten to bring the water with which to wash down our lunch.

I spent a lot of time at Bungle Gully rolling bracken, controlling rabbits in awkward places, trying to establish good perennial pasture. At one stage I had the whole 400 acres rabbit-free but my efforts at pasture establishment were indifferent, despite the use of trace elements on the sandy soil. The area seven miles south of *Glenspean* had a lower rainfall. I was the drover who negotiated mobs of sheep past the various landowners on the trips north or south. With its sale, all that part of my life has just disappeared – thank goodness.

Rabbiting is another activity that I have not been sorry to relinquish. Of late the Calici virus has been very effective: rabbits just disappeared. However, Sandy, with Julie in the driver's seat, continues to remove warrens and do some 1080 poisoning lest the virus loses its virulence. Newer, more virulent strains are disseminated by the authorities as they are released.

8 Using my agricultural degree

In February 1952, I saw Mum and Dad off on the *Himalaya* for a ten-month trip around the world, mainly in the UK. So there I was alone, a complete neophyte, with no idea at all about how to manage four full-time staff. John eventually came home from his honeymoon. After brief service in the air force towards the close of the war and with seven years of practical experience under his belt he became the managing partner.

As a schoolboy, and then as a university student, my vacation role had been to muster sheep from distant parts of the farm, even from across the river over the ford concreted by John and Emmett. I would take my mob past other sheep in each of the succession of paddocks between there and home.

In addition to mustering, at the end of each university year, I was called upon to help with the hay making. In November 1951, I completed my four-year agricultural science degree at Melbourne University.

Except for lectures on animal diseases, the course did not include animal husbandry, and for that matter there was not a lot about experimental agriculture at all, anywhere. I came home and helped cart in the pasture hay, et cetera, as usual.

Early in 1952 I went to the Dean, Professor Sir Samuel Wadham, asking how I could utilise my agricultural degree. Little me: son of my innovative father, that was my smallest of innovations. Sam put me in touch with the faculty's first, recently appointed, senior lecturer in animal husbandry, Terence J. Robinson (later Professor at Sydney), and that was how the Glenspean Experiment was born.

Cattleman Neil and Prof Terry Robinson

Flowing from the Glenspean Experiment, and following in our parents' innovative footsteps, John Cameron and I were the first cattle breeders in

Victoria and possibly in Australia to select Aberdeen Angus bulls on their ability to put on beef, when others were guided by what looked pleasing to the eye in the show ring. For that matter the breeders of dairy bulls were guided, not by the relative performance of their female progeny in producing milk, but by some ill-defined image of what made a good bull.

Likewise the breeders of Dorset Horn rams were influenced by not much more than the shape of the horns.

(Note: performance testing dairy bulls requires statistical comparisons to be made of the ability of one sire's *progeny* to produce milk to that of other sires' progeny, whereas the ability of sires to put beef or mutton on their progeny is moderately heritable and can be measured directly from their own growth rate.) My father, also Neil, was the great innovator, unlike me in my youth. His leadership in farming practice was not widely recognised even in the local community until just a few years before his death. Dad's ability to think around the square was shared by his brothers, who built up their own substantial farms, beginning pre-war when the oldest four of the six brothers Cameron paired off on small holdings to the north of Melbourne to hand-milk a few cows, which they were wont to name after their girlfriends. Post-war, the five surviving brothers drew upon such scientific expertise as was available in those days as they and their families spread themselves across three states. Between them and some fellow thinkers they set the pace amongst farmers who were willing to 'take the blinkers off', and leave tradition behind. My father Neil Wilson Cameron did just that.

That introduces a number of other extraordinary features of the Glenspean Experiment. The first was that the property existed at all in a way that was conducive to its conduct. *Glenspean* had a sizeable herd of black poll cows and calves, incidentally making it easy to recognise our cattle from the neighbours'. For the practical conduct of the experiment, horns were a no-no. At that time the alternative breeds in Australia, Hereford and Shorthorn, were all horned. One of Dad's innovations was to calve his cows down in the autumn rather than the spring, which was the traditional practice. Dad had always run cattle in addition to sheep. At the beginning they were used to eat down the tussocks.

In January 1950 it is recorded that he had 147 cows in calf, six bulls, 66 yearlings and 185 weaned calves, not forgetting the three milking cows nor the 12 stud Angus cattle that he had just purchased, so it was easy to find cows for the first mating in July 1953 for an April 1954 calving.

Only the mature cows were used, which, together with three of the bulls, formed the experimental herd.

The subject of the experiment was thus the Angus herd on *Glenspean*, now divided into spring and autumn calvers. To obtain a similarity of age, the first flush of calves to be 'dropped' in each season was used.

The object of the experiment was to make beef production more efficient, which involved improving the growth rate of calves and weaners.

Due to Dad, *Glenspean* also had improved pastures that might be expected to give measurable growth rates in young cattle.

Probably inspired by his relative, Hugh Trumble of the Waite Institute, South Australia, Dad determined to 'make two blades of grass grow where one grew before'. Now *Glenspean* was in shape rather like a squashed banana, the narrow part just two paddocks wide in the middle, with bulges at each end. By the outbreak of the Second World War in 1939, despite the Great Depression of 1929, or perhaps because of the economic necessity it caused, Dad had sown to improved pasture most of the narrow part and some of the home and distal bulges of the oddly shaped *Glenspean*. He sowed the only existing cultivars of subterranean clover and perennial ryegrass, each at 2 lbs per acre ('Mount Barker' and 'Victorian' respectively), top dressed every autumn with one cwt (112 lb) or a bag (186 lb) of superphosphate per acre, depending upon finances. Few Victorian pastures had been so 'improved' before post Second World War the soldier settlers arrived around Meredith in 1952–53, by which time the Glenspean Experiment was well underway. Whereas the native pastures of wallaby grass and silver tussock had a carrying capacity in Dad's words 'of two Merino wethers to 3 acres, badly', his improved pastures carried two crossbred ewes per acre, a sixfold increase in productivity. Unfortunately, by the time of the Experiment, the experimental paddocks had been sown for twenty years and suffered from some clover decline and an influx of weeds such as winter grass (*Poa annua*) and silver grass (Vulpia spp).

I took pasture samples regularly for analysis of their protein and energy contents. Later on we found a potassium deficiency to be the chief cause of the clover decline although the build-up in soil nitrogen had reduced clover's competitive advantage over the weeds.

Polled cattle, improved pastures: our next requirement was scales.

From my viewpoint as a partner in Glenspean Partners, 'our' big innovation was that the experiment was almost entirely self-funded rather than being a project for an institution. I think the scales may have been paid for by the university. We installed the beam balance scales in a concrete-lined pit, with space underneath for the beams and knife edges supporting the 6 foot x 8 foot platform carrying a pen with gates at each end. Old-fashioned by 2018 standards, the scales had weights representing 500 and 100 pounds on notched beams and a slider for weights 1–100 lb. I remember the owner of the Colonial Scales Company setting it up. When he had got the scales levelled and balanced exactly, he threw a packet of cigarettes onto them, and the scale beam went up! Not bad, eh? Designed to weigh 3600 lb, they were accurate to 2 ounces!

Left: scale house and scales; entry race from walk-through cattle crush on left
Right: triple beam scales with notebook in place

To our existing cattle yards we added, with practical help from Terry, a forcing pen, lead-up race to the walk-through crush, a photography bay and the scales themselves protected from the rain by a shed. The walk-through crush was one of John's inventions. With plenty of imagination and having acquired the engineering skills to put his ideas into practice, John is a true Cameron of *Glenspean*. He subsequently rebuilt the weighing pen to make it more usable, making a narrow pen that animals could not dance around. He gave it sliding gates at each end.

To weigh the calves at birth required a utility, a set of bathroom scales and some agility. I would drive up to a freshly born calf, wait until the dam's attention was temporally diverted, grab the calf and jump with it into the low back of the Vanguard ute, where I stood on the scales, subtracting my own weight, tagging the calf and recording the number of its dam. I have seen a cow so frantic to protect her own calf that she stood on it, but in the long run we all survived! The next requirement was notebooks narrow enough to fit into a pocket and of sufficient length for 36 entries, with columns for successive weighings. Waller & Chester of Ballarat obliged. Oh! and pencils too.

This experiment had a factorial design that was unique, especially so when applied to large animal experimentation. I do not know if Terry

Robinson's method has ever been repeated. His series of firsts had begun with that smallest of ideas – mine – when I had asked the Dean of Agriculture how, buried on the farm, I might make use of my degree. It worked out brilliantly.

Terry Robinson's world-famous Glenspean Experiment 1953–58

Terry's aim was to grow, as quickly as possible, animals of suitable weight and grade for chiller beef, juicy and tender.

Up until the time of the Glenspean Experiment, there seemed to have been virtually no experimental animal agriculture in Victoria, with the advice tended by Department of Agriculture officers formulated by experience rather than by experiment. The Meat Board had financed the installation of two sets of cattle scales, one in the north-east of Victoria and one in the Western District, but they were only used to record the weight changes in weaned cattle up until slaughter, which may have been at three or four years of age. With the annual weight loss in winter having to be recouped before advantage could be taken of the spring flush, it took several years for cattle to reach slaughter weight, which led to tough beef made barely tolerable by an excess of fat on the carcass.

Terry Robinson's experimental method contrasted with that of the Meat Board, or rather its lack of one! Our experiment was famous mostly for putting experimental animal husbandry on the map, then for its experimental design, and least of all for what we discovered. It was famous also for having been done at all, leastwise for having been done over an extended period with large animals – to wit cattle. Terry Robinson's greatest innovation to Victorian animal husbandry, his Brilliant Idea, was to inject into it the notion of experimentation. His dictum was: 'You must apply a treatment, compared to a control: in that way you will always get an answer, even if it is negative.' As with climate, agriculture being subject to random variation, any experimentation needs to be replicated. It is so variable that observations must be repeated and subjected to statistical analysis to assess the probability of their being due to chance. Testing the response to a particular treatment of growth rate in cattle requires many animals, some to be the 'controls' with no treatment applied.

Terry's second Brilliant Idea was to apply not just one but five different treatments in which the group as a whole provided the 'control'.

To study the various factors influencing the weight gain of calves he designed a 3 x 3 x 2 x 2 x 2 factorial experiment, with the five factors involving 72 calves. For any one factor the other factors provide the numbers for statistical analysis.

Breeding value

Calves were sired by three bulls of varying conformational beauty according to current show standards, one a registered stud bull, one average, the third a real scrubber. The bull groups had coloured labels on their neck chains (blue, red and white), in addition to the calves' ear tags.

The males' labels carried odd numbers, the females' even.

Nutrition

For comparision, calves and their mothers were given three levels of supplementation to grazing comprising:
- Zero: grazing pasture only
- Hay fed when pasture was limited in amount or quality
- Hay plus grain creep-fed to calves in off-seasons of pasture; also grazing oat crops

Other factors

There were comparisons between two sexes, male castrates and spayed females; two seasons of birth, April or September; and two years of birth, 1954 and 1955. For statistical purposes the controls were provided by the numbers, e.g. 36 steers compared to 36 heifers, likewise 24 animals in each feeding group compared to the others, and so on. In autumn 1955, 12 extra male calves were carried through and chipped with 'stilboestrol' oestrogenic implants to test its effect on growth rate. It was a brave experiment based on the assumption that all the animals would survive.

We did lose one, but we had spares that we could plug into the gap with some statistical adjustment.

First there was the mating in three separate bull groups. Numbers were sufficient to allow 30-odd cows to a bull group in each of the four mating seasons, spring and autumn 1953 and 1954, enabling us to utilise the first few calves born, so the experimental calves were of equivalent ages. Then there was allocation into feeding groups, identified by the metal tags around their necks: round, triangular and square — typical Robinson humour.

Once their calves were weaned, the cows were returned to the main herd. They were reallocated into bull and feeding groups for the second year calving. Mating in bull groups for the second year calvings was awkward as it cut across the feeding groups. I quite frankly don't remember how we did it.

On 1 June 1954 and again in 1955, numbers being sufficient, we divided up the autumn-born ones into their feeding groups of 12 cows with their calves, which were rotated weekly around five small paddocks of 9–10 hectares each.

The paddocks set aside for the experiment, the '77 Acre' and the '55 Acre', were subdivided by primitive electric fencing into three and two respectively.

The idea of the rotation was to remove any influences on performance due to differences in productivity of the pasture, in addition to the differences in hunger of the feeding groups, leaving more, or less, for following groups to eat. That was an inherent fault in our set-up due to the limitations of *Glenspean*, it not being an experiment station. On such an experiment station the different feeding groups could be allocated to different areas and the whole experiment repeated over a number of years, the various groups being rotated around the station. For financial and logistical reasons, that was not possible on *Glenspean*. With the rotation one would expect the 'zero' unsupplemented cows to be hungrier than those whose diet was supplemented with hay, or hay plus grain. Particularly with the slow pasture growth of the winter season, they might be expected to graze more closely, leaving relatively less for the supplemented groups.

Glenspean Experiment heifer awaiting photography to measure growth

When animal performance was measured by weight gain, its sensitivity to the amount of pasture available is shown by the following records I made: in winter when the sub-clover measured 5/8 inch high and the ryegrass 1 1/2 inches the 'zero' cows with calves at foot lost weight: when clover height was 3/4 inch and ryegrass 2 inches those unsupplemented cows gained weight.

At the end of spring the April-born calves were weaned and reallocated to their feeding groups, whilst those September-born and still on their mothers were allocated to their feeding groups for the first time from when their dams' milk supply began to fail as the pasture matured in the experimental paddocks until weaning in late autumn, when they were run together with the autumn-born weaners of the same feeding groups, in separate paddocks to the experimental paddocks now set aside for the 1955 autumn calvers. The second year was different because we had good green feed from late summer rain.

Sheep were used to control rank pasture in all years.

Those receiving 'square meals' were fed whatever was required to keep them going, including an oat crop. Creep-feeding with oat grain made little difference to growth rates and the old codger who sold us the oats was most indignant that their quality should the questioned. I now realise – Terry Robinson should have – that oat grain is not of high enough energy content to provide a useful supplement, compared to barley or wheat. In the spring flush all cattle were depastured together.

The 'triangular' groups were fed hay in winter and the 'round' group nothing. Autumn-born calves were carried through as weaners until 21 months of age; spring-born ones to 27 months because they were not big enough to be slaughtered at 15 months, the finale of the experiment being the slaughter in January 1958 of those born in September 1955.

The operator was me, the man-on-the-ground who turned my first little idea into an extensive reality. I did have lots of help from John and the farm staff, but the weekly rotation, the mating in bull groups, the surveillance of the calving cows, the identification of the calves with neck chains, the establishment of groups and their supplementation, the maintenance of the fencing and mustering on horseback were my responsibility.

Rotating cows and autumn-born calves around the five paddocks was done on horseback as was mustering them to the cattle yards for the monthly assessment. The yards were on the hill three miles from home along the Mt Mercer Road. Every month Terry would bring down one or two assistants to help photograph, weigh and measure the animals. They would drive down the Princes Highway towards Geelong and turn up the Midland Highway, and as they rose up Bell Post Hill on the outskirts of Geelong it would start to rain – 'horizontal rain' Terry called it. Once, when Terry was measuring the cannon bone of a calf, it kicked him.

'—' (expletive)!

Me facetiously, 'Did it kick you?'

'—' explosion!

Growing cattle store energy

In each case slaughtering was done in mid to late January when the carcases were weighed hot, the caul fat also. Whole carcases were measured and photographed, then quartered at the 11th–12th rib. The eye-muscles were measured for length and width. They were photographed and samples taken for analysis of bone, muscle and fat ratios by Terry Robinson.

The significance of the fat to muscle and bone ratio is that it takes ten times as much energy to store a pound of fat as against a pound of bone and muscle so that weight gain does not necessarily indicate how well the cattle are doing. Over lunch one day during the monthly assessment I commented to Terry, 'It's a pity that you cannot measure the electrical conductivity of a cow, to determine its fat content.'

Ex-naval officer Terry nearly jumped out of his skin: 'ASDIC'! (sonar used to detect enemy submarines).

At Terry's behest the Melbourne University physics department began constructing a sonar probe to measure the fat thickness on the live animal from which to deduce its fat content: but the probe was never completed because in the sonar stakes the Yanks beat us to the draw. Oh well! In the meantime we had to measure it in the abattoir.

At the conclusion of the experiment, Melbourne University's Dr PJ Claringbold analysed the weight gain results using the university's first-ever computer, 'Siliac', which occupied about the entire basement of his Physics building.

Viewing the data in a slightly different way – final weight at slaughter as against weight gain from the first weighing – I actually did an analysis of variance manually, before giving up entirely. When I attempted to write up the experiment for a master's degree there were very few papers in the English-language literature about factors affecting growth rate in cattle (or wool growth in sheep for that matter). Hammond in the UK had quantified the muscle–bone content of carcasses as affected by lifetime growth patterns. One renowned experiment in California compared the growth of calves supplemented with cottonseed meal, either in the greenfeed season, or in the dry season. Of course the latter did better! I had a barney about it with my supervisor, one of Terry's post-docs, maintaining that it should have been conducted as a 2 x 2 experiment, with 'supplement' / 'no supplement' comparisons being made in each of the dry and greenfeed seasons. I never did complete my thesis, but by then I had a growing family to nurture and a farm to run.

Terry wrote up the results in four papers over our joint names, the first published in the *Australian Journal of Agricultural Research* (*AJAR* vol. 11, no 6, 1960). He put me as first author of one, which I did not think I deserved: 'The effects of a number of management practices on chiller beef production in Victoria', NG Cameron and TJ Robinson, *Australian Journal of Agricultural Research* 13(3) 448–460, published 1962.)

Results

The squares were significantly heavier than those other two groups supplemented with only hay or with nothing (highly significant $P<0.001$, i.e. there was less than one chance in 1000 of the results being due to chance), with little between those two lesser feeding regimes. In the spring, the 'nothings' made significant compensatory growth, but never caught up to the pampered ones.

The autumn-born calves were significantly heavier ($p<0.001$) at weaning than the spring-born ones.

Rex McKellar, manager of the Ballarat Victorian Inland Meat Works, was most concerned that the fat cover on the carcasses would be insufficient to provide tender meat, but Terry Robinson was adamant that by being killed at a young age they would yield succulent beef. It was quite so, the more so because the cattle had been transported straight from the paddock to the abattoir. So if we achieved nothing else we provided an impetus to the industry for selling at a younger age.

As a consequence of the experiment we took to selecting our commercial herd on weight gain, but since no weight-gain-tested bulls were available for purchase, we mainly used our own bulls and so ended up with an inbreeding problem. Visual symptoms included bent penises, making bulls unable to serve a cow. Then there was curve-claw too, which nowadays breeders may be tempted to cure with an angle grinder. Of course, inbreeding also meant that our bull-breeding herd as a whole was not as productive as it would otherwise have been, but we had no actual measure of that. We did know that in earlier days when we retained Dad's Angus stud, its growth rates were less than in the commercial herd when we used purchased bulls, before we took to breeding our own. To cope with the inbreeding problem, in at least one year we kept all male calves entire so that they could be selected on weight gain at the end of their second spring. The idea was to provide at least eight new sires for each generation to reduce the impact of inbreeding. A new problem then arose: the bulls tended to fight a lot amongst themselves.

In that era we spent a lot of time measuring testicular size and an individual bull's ability to repeatedly serve a cow restrained in a crush.

Not being brash entrepreneurs we could only command a price for bulls double their meat value, which was not sustainable, so eventually we gave up selling them. It had not helped that our clients always wanted to come on a Sunday.

We followed the breeding path with our sheep as well. We bred the horns off our Dorset Horn flock, if my memory serves me aright by crossing and back crossing with Southdowns, which are polled.

An interesting example of inbreeding in sheep is the Border Leicester breed. I purchased a small flock with the view to introducing the Booroola gene, which was supposed to increase the number of twin births. My new flock only had a lambing percentage of 70 per cent, similar to that of Merinos; however, when the two breeds were crossed to produce the Border Leicester – Merino ewes as prime lamb mothers, the lambing percentage jumped to about 130 per cent, depending upon the body condition of the ewes at mating and again in late pregnancy.

I joined a group sponsored by the University of New South Wales whose intent was to breed the black points off the Suffolk breed, supposedly superior to Poll Dorsets when used as terminal sires in prime-lamb production.

When Sandy and Julie began to milk sheep it turned out that the White Suffolks were better milkers than the Border Leicester – Merino ewes with which we began and were absorbed into the milking flock. These breeding programs of mine went by the board. They became of nuisance value only.

From a personal viewpoint, accepting on-the-ground responsibility for running the Glenspean Experiment was a vital first step in my ascent to manhood after the very shaky start, which I described in an earlier chapter. Then I decided to woo June with the wonderful, ongoing result of family, giving continuity to the Camerons of *Glenspean*.

Completion of my ascent to manhood came from the third, most vital step taken not by me but by the Spirit of Jesus I suppose it was, who gently pursued me until I eventually said 'Yes' to He whom I had rejected at age ten. That occurred after life-giving preliminaries such as learning to ski and to sail, and putting myself through university. It is a contract with the Spirit of Jesus by which I share myself with others, both monetarily and otherwise. It provides me with the initiative and innovation in my dealings with farm, church and society. That is the theme underlying the final chapters of this book. By my work with Prof Terry Robinson on *Glenspean*, I, Neil Gordon Cameron, became widely known amongst the scientific fraternity of the livestock industry. My tongue had been loosened and I used it.

9 Fireman Neil

In my first experience of fire-fighting, we rushed up to a fire on *Narmbool* and were roundly abused for not leaving the utility on black ground. We were superfluous anyway because the fire had been extinguished by the local fire brigade truck. So we needed a fire unit of our own.

The first fire unit owned by Glenspean Partners was a cylindrical 100-gallon galvanised iron tank and pump lying in the bed of John's utility. With it he put out a small fire coming up from the river below Camberley before anyone else arrived.

By the time of the 1967 fire I had a 200-gallon tank on the back of a four-ton truck with the pump attached to a suction hose which could be dropped through the fill-opening on the top of the tank, or into a dam to fill it with water. From the petrol-driven centrifugal pump, with a hand primer to suck water into it, water was delivered into a couple of hoses fitted with taps and sprays. Not so long before that fire I had been to one in scattered timber below Bolte's farm. Burning up from the river the fire had been extinguished but to ensure that it did not break out again it was necessary to look for any wisps of smoke near the edge and extinguish their source. Other fire-fighters there were spraying the blackened surface whenever they saw one; but I discovered that it was best to direct a jet of water vertically downwards in the vicinity of the wisp of smoke when hotspots could be found and extinguished with very satisfying explosions like miniature volcanos.

Black Wednesday, 22 February 1967

That date is indelibly imprinted upon my memory, augmented from several pages of my diary entries made at the time. My story makes reference to just about all my other activities around that time both on and off the farm. It also draws in many of my neighbours.

In the third week of February 1967 we were shearing. John was in the shearing shed classing the clip. He had come over from Camberley in his fire unit, leaving mine with his assistant Harold McKenzie in case of fire over there. His one-man fire unit comprised a Bedford 30-hundredweight flat-top equipped with just another of his ingenious devices. One hose from the pump connected via a tap in the cabin to a piece of half-inch galvanised pipe leading down through the firewall to just above the front bumper bar where it had a 40° bend and flared end which, with the water turned on, created a slightly fan-shaped jet of water directed to fight a fire *ahead* of the vehicle either to left or right. Half a century later volunteer brigades are provided by the Country Fire Authority (CFA) with trucks having a directional front jet.

※※※

6.00 a.m.: out mustering early.

8.00 a.m.: June having havered about going to Geelong left instructions with Margaret Innes to ring her at Shannon Park if there was a fire anywhere.

11.45 a.m.: fire starts from an incinerator at the Lal Lal pub (officially children and matches)

Jeff Cooper arrives with kitchen cabinets. It was just on lunchtime on the third successive day of over 100°F. John Vanstan began unloading the cabinets for our renovated kitchen.

11.50 a.m.: John got a telephone call from our builder Alec McDonald with whom he lived and worked, Alec having stayed home on account of fire risk: 'Come home there is a fire to the north of our place.' That meant it was to the north of us as well!

Telephoned Beb at *Camberley* on the party line to send Harold McKenzie as the fire is on his family farm. He had been left with my truck as John had brought his unit to the shearing shed.

※※※

I alerted John in the shearing shed, saw the bearing of the fire, almost north, fractionally east – just a puff of white smoke; ran back to the shed just as they were knocking off. 'All hands!' Rouseabout Pat and learner-shearer Brian were both endangered. Scramble! Shearers Frank and Leo manned the Meredith Brigade truck.

I grabbed a few things including a plate of lunch and a gallon of water, but forgetting my transistor and my green fire brigade lieutenant's hat I sped with ute to meet Harold on the Mount Mercer/Elaine Road.

Mistake 1: I should have had Keith Innes drive me to meet Harold so that he would be left with the ute.

We crossed the Midland Highway several miles north of Elaine and continuing north a mile or two met the fire burning south in open country. We were on our own with no other units around to back us up. The head of the fire was running south at maybe six or seven miles an hour in grass only a hand-span high. Useless trying to stop its head even though its flames were only thigh-high; so we turned around and went further back towards the source of the fire, then turning again, driving in the same direction that the fire was travelling, started to extinguish its right flank. We had not gone back far enough and we kept being outflanked by secondary fronts which had to be extinguished. You might be able to imagine holding a piece of honeycomb and trying to stop it dribbling onto the floor without first making sure that its edges were contained. The fire was a bit like that, with 'dribbles' or secondary fronts coming down from near the fire's origin at Lal Lal.

It turned out that there was a salt pan not far from the edge of the fire and with a coordinated attack by several additional trucks it would have been possible for them to actually light a fire, back-burning towards the main one, extinguishing the southern edge as they went. It would have prevented those secondary fronts coming down and enabled concentration on the main one.

Mistake 2: I should have gone in convoy with John. With two of us one could have dealt with the secondary fronts; but I was too busy doing what had to be done to think about that then. As it was we had no means of communication with each other.

Lesson A: secure your flanks!

We had turned around at least once to extinguish the secondary fronts coming down beside the main fire; but eventually gave up around 2.45 or 3.00 p.m. and headed for home, nearly hitting a police car and hearing from it that '(Willie) Grant's house is gone' (on the Meredith Mount Mercer Road). We had had no other information about the fire.

By this time the head of the fire had well and truly crossed the highway. We heard later that John Cameron with his light unit had put out two 'spots' that crossed it but could not reach a third because of the drain – anyway it swept across the highway nearer to Elaine. (The highway runs in a nor/nor-west–sou/sou-east direction, diagonally across the path of the fire.)

Below is a litany of reports, which I may have heard on my transistor, had I taken it with me (as brigade radio operator I had an army disposals set that was too cumbersome to take in the truck; in the hard-hat brigade, as a lieutenant I toted a green hat, as radio operator a yellow one. I'm not too sure which colour was the more appropriate.)

With a radio I may have heard the following:

∗∗∗

1.15 p.m. Ted Tansley signalled from Murphy's lane 'fire out of control'
1.55 p.m. 'heard 'Harts' house safe'
2.00 p.m. 'fire at Grants'
2.03 p.m. 'trying to hold fire at Boundary Road'
2.20 p.m. 'fire jumped Boundary Road'
2:25 p.m. 'six trucks ordered to Grants'
2:40 p.m. 'Bamganie Forest ablaze [from spots]'
2.45 p.m. 'fire at Crows – south of Hugh McColl's'
3.55 p.m. 'fire approaching Stewarts' [Henderson Road]'
4.10 p.m. 'fire past Stewarts'

∗∗∗

The above litany of reports in my diary showed later that we had been well and truly left behind by the fire. Incidentally, Jack Ridd on the eastern edge of the fire had a narrow escape. The fire swept past just west of his house. A 'spot' started a second fire just south-east of his house by 300 yards.

There being no direct route home we headed west dropping off Tommy Hart at his front gate. He had been helping Jack Ridd on the east flank of the fire. On the west flank it had burnt under Tommy's relatively new 1952 soldier settlement house without igniting it. The fire had swept through some time before, leaving unburnt some bales of hay, his house and the majority of his trees on its west side. Where it had burnt under his house it wasn't even still smouldering. There was nothing to be said to Tommy – we just drove on.

Ted Tansley wanted us to help him at Clarrie Smith's on the north side of the road where another front was burning through plantations west of his house – however, it was anxiety over our own situation which made me press on towards home.

I later wrote in my diary that my reason for staying in Wells' north of the highway so long was to protect the flank of those fighting fire further south: that I trusted that masses of trucks would be in attendance further down. My memory is that this was not really so; but rather that we stayed because there was fire there to be fought and fight it we did, without thoughts for what was going on elsewhere – anyway we only saw two red brigade trucks all day and perhaps a dozen private units.

Some or all of the Lawaluk and Grenville brigades west of the Leigh River were told to stay there until 3.30 p.m., when it was too late.

At the place where the Shire bore and tanks now are, we waited for the fire to approach as it snaked down through a paddock previously mown for hay. As we learned later the approaching fire was kept narrow by John Cameron and Cec Barber patrolling its flank. We tried to stop it jumping that angled section of Bamganie Road which heads in a south-easterly direction. Unfortunately there was a very small triangular pine plantation on the north side of the road which burnt in about a minute, giving us a scorching and preventing us from holding the fire. It jumped Bamganie Road, so was on the left of Bamganie Road where it turns south. By this time we had been joined by brigade captain John Nolan with his farm truck. With John Nolan's and our pumps not working well, it took half a mile to run that spit of fire into Bamganie Road as it continues south. I immediately turned back to patrol what we had put out, to the north of the angled section of Bamganie Road. There were a few flames that were burning back towards Musgrove's which I decided to put out; but whilst manoeuvring to do so we spotted a foot of flame just behind us. By the time we had turned again (perhaps 30 seconds) it was away again. Our pump pressure was worse and we just couldn't keep up. John Nolan had gone to fill up at Morrison's dam and we had to do the same. It was this second blaze that burnt the surrounds of Stan Morrison's house, burning half his pine plantations too. John Nolan stopped it crossing Cameron Road. He later told me that they had run out of water and resorted to beating the last flames out with their hats. According to him a nearby truck of professional fireman from the city refused to risk themselves by helping.

The main head of the fire had crossed east-west Boundary Road where Betty Payne's soldier settlement farm on the north side and George Lloyd's on the south side had been directly in its path. Betty's house and the plantations immediately surrounding it were saved by Joe East, who had gone there with the specific intention of protecting the widow. I do not remember how George Lloyd on the south side of the road fared, badly I suspect except that he saved his house.

The Shire engineer's fire protection plan had worked to the extent that whilst it did not stop the head of the fire it narrowed it down, cutting off those secondary fronts coming down both sides.

John Nolan and I refilled our trucks at Stan Morrison's house dam. It was so hot that I filled my canvas hat with water and put it back on my head. John discovered that the bag which he had put in the fill-opening at the top of his fire tank had dropped down and was obstructing the suction hose. For my part, the pump was not working efficiently enough to suck water out of the dam, because the previous day Keith Innes had had the face of the pump off to clear away some debris which had been sucked into it and was only beginning to tighten the suction hose to it when it came on lunchtime and the shearers were coming to his place for lunch. On the day of the fire the connection gradually came loose again.

Lesson B: you think you are ready; but have you tested everything again? Of course neither he, nor John nor I had any idea that there was to be a fire.

By the action of the two Johns the fire had been contained to the east side of Bamganie Road as it burnt south but jumped westwards across it again where there are some trees half-way down towards the Woodburn school. It burnt through half the school grounds but on that flank was stopped again by Mount Mercer Road. Margaret Innes had rung June working at Shannon Park to tell her of the fire, whereupon June rushed home. By the time that the fire burned through the school grounds June had already collected the whole school fraternity of eleven children plus extras, stacking them into the car (and head teacher Greg Lee's car too?), taking them back to *Glenspean* out of the path of the fire (with the Grundy children saying in the bath, 'We'll get a belting for this. Our dad doesn't let us stay overnight').

At a point where there was another patch of timber 500 metres south of the Woodburn school it again jumped Bamganie Road from east side to west side. That secondary front went southwards, jumped Coopers Road and destroyed Arthur Elliott's old house and Gordon Elliott's. As the fire passed, Jack Stevenson rushed up from Lethbridge and with a wet bag saved Drew and Christine's newly purchased house. Dick Ratcliffe-Smith on the east side of the road, between it and the Bamganie forest, lost all his sheep feed and 400 sheep were burnt – the remainder were on bare ground. His pines around the house were scorched to the top like those at the school and at Stevenson's, but saved by a squirt from someone. Jim Musgrove's old house south of Dick's went 'whoosh'. That's what fire does.

Well, after putting out two or three loads of water at Stan Morrison's, tightening the suction hose and towing two trucks to start them, then some humming and hawing as to what to do next, we set off down the Bamganie Road in pursuit of the fire, seeing no one. We turned off down Bath's Lane on the west side of Bamganie Road but after our previous experience, turned back off it to get behind the flank. With two other trucks we put out the flank as far as Henderson Road where Bamganie Road terminates.

Then as dusk fell around 7.00 p.m. a light south-westerly came in and we retired from the scene. I arranged for all night patrol by shifts of volunteers, took firefighter Bert and another back to Elaine and called to see Betty Payne. They had saved her house, its plantations and half the sheep. What an experience for a widow!

We have been talking about the west flank of the fire: on its east flank, opposite to us George Colclough had all his sheep on a bare paddock, saving them and his house. Driving past afterwards you could see amongst the black several strips of unburnt grass where fire-fighters had tried to put out the blaze, but lacked support, and its speed made their efforts ineffective. The same sort of thing happened at the 1969 Lethbridge Fire as it headed eastwards towards the highway.

All this time we had been unaware that Captain Don Wallace had led eight or nine trucks of his Teesdale Brigade north up Bamganie Road in search of the fire; but coming to Meredith Mercer Road and realising that the head of the fire had passed he turned around and led them south again as far as Henderson Road, when he turned left and proceeded to do a back-burn. (I was not there nor read a description of what took place, so had to imagine it.) With the trucks providing a water break they lit along the edge of the road, then, with firelighters trailing, men ran northwards in the face of the oncoming blaze to break it up.

Don later commented publicly that those guys deserved a VC.

Aftermath

Thursday 23 February 1967

(With explanatory additions, my diary for the following week reads like this:)
No shearing! Hot, another century; variable light winds

After getting to bed at 11.00 p.m., a call at 3.00 a.m. from Deputy Group Officer Ted Tansley and Brigade Captain John Nolan asked me to take charge of a section of CFA tankers, at the school at 5.30, which was no hardship as I had arranged to patrol from four anyway. Well! The sight of it. Three CFA small town tankers, very modern, from Drouin, Longwarry and Loch, manned by volunteer fireman, and a seven-tonner with professional firemen from Springvale, to patrol two and a half miles from Bath's Lane to Bamganie. As it turned out we needed them to put out the large number of timber logs and stumps. I finally turned it over to the all-nighters at 6.30 p.m. There was light enough left to see Chief Veterinarian Dave Wishart supervising the slaughter of 67 of Gordon Elliott's sheep: they didn't want them to be shot. Dusk was gathering. Mrs Elliott, normally the one to be contended with, had retired from the scene.

Friday 24 February

No shearing, hot again; moderate-fresh northerly wind. Spent all day patrolling 7.00 a.m. to 5.00 p.m. Periodically, the professional fireman in the Ballarat (or was it Geelong) replacement seven-tonner would report 'It's black out'. Donning my green hat I would find some more hotspots for them. 'Okay, you're the boss; we can't go until you release us', they'd say, no doubt cursing that country know-all.

With two Deputy Group Officers, David Brown and Ted Tansley, Captain John Nolan and Assistant Chief Officer Chester Nevett, as the lieutenant in charge of that flank, I had to decide how to make the river safe so I made them take a seven-tonner over a maze of rocks to put out the two remaining hotspots. Someone went down and threw a log or two into the river. More mutterings but infinitely preferable to burning a break on the far, west bank of the river. At 4.00 p.m. welcome rain fell and at last I gave the order releasing my brave hard-working beauties. And the volunteer teams from Gippsland could go home too!

Saturday 25 February

Recovering a bit.

With fellow councillor Harold Cook I spent the afternoon doing a systematic interview with the 25 victims of the fire. The Shire Presidents of Bannockburn, Buninyong and Leigh Shires had initiated an appeal for hay, with a government grant of $5000 to be used for cartage. We decided on a formula for distributing the gift hay: stock surviving less acreage of usable grass by a factor of 3 to 5 equals the number of a bales of hay needing to be allocated to each victim. We thought, 'That should ensure that initially everyone gets enough for three weeks; it could well rain in that time.' (It didn't: it was the beginning of the 1967-68 drought, probably the worst over the hundred years before the time of writing.) This gave a figure of 16,000 surviving sheep in the burned areas of the three shires needing to be fed.

Some statistics. (Stock losses: Buninyong Shire 10,500; Bannockburn Shire 6500; Leigh Shire 3000; 200 miles of fence, mainly old; and 25,000 acres of grass and forest. There were 32,000 surviving sheep in the whole Bannockburn Shire, including the areas not ravaged by the fire.)

Sunday 26 February

Up betimes and erected direction signs to and nameplates on the various farms to which hay might be sent. A few public spirited chaps, including Harry Mohr and Arno Hall, had already left a ute-load on each farm – a wonderful morale booster. Patrolling was still going on with occasional trees flaring near the edge of the burn. Before church I had time to spend 15 minutes with our youth group. Of the church people, I noted in my diary that 'Peace-loving Campbell Lamb and quiet Tom Fishburn had certainly done their bit on the Wednesday. They couldn't light a fire save themselves at Tom Harts (whatever that referred to); but Tom had well and truly squirted an inquisitive photographer who got in their way'.

Monday 27 February

Shearing resumes.

I mustered the crossbred ewes brought in from South River – four died. On Tuesday, a fifth ewe died – I rushed it up to the Ballarat vet, Len Fulton, who seeing no symptoms asked, 'What is the problem?' I replied, 'She's short of breath', which turned out to be too true to be funny. She had viral pneumonia, a disease new to us. Len's advice was to leave them alone so we stopped shearing.

Friday 3 March

Dipped using Vetamax Plus (plus sulphur for itch-mite) (spray dip)

Put off vet Fulton again as the road past the cattle yards has just been primed for sealing. Fulton was to have come to pregnancy test the cows last Wednesday, the day of the fire.

The Meredith Memorial Hall was packed for the final wrap-up of the fire. I vented my ire against those who had on the day of the fire parked a whole lot of visiting brigade fire trucks in a vacant township block instead of sending them out in groups to fight the fire. Mind you, it is not always easy to know where to start (which I discovered two years later when attending the 1969 Lethbridge Fire for which the only diary entry I have is that it occurred again on a Wednesday).

Lethbridge Fire, Wednesday 8 January 1969

The fire had been started by a grader on the Meredith Shelford Road (under jurisdiction I am pleased to say of the Leigh Shire, not my own). It had burnt south as far as the Bannockburn–Shelford Road; then was picked up by a violent westerly change which brought it back crossing the Midland Highway south of Lethbridge before being slowed by the steep banks going down to the Moorabool river. Hot air rises so fires don't burn very well downhill.

When called out by the brigade I jumped in the truck; young Terry Innes and a couple of mates jumped on the back – it was to be the experience of their life. We headed south down the Midland Highway, turning into Lethbridge. Crossing the railway line, there was old Max Donaldson directing traffic to the left, south-westward, but we hadn't gone far before being confronted by a wall of smoke; so I fled and instead went westward along Tall Tree Road. We inconsiderately cut a fence to gain access to the fire on Tall Tree when we could have gone a little further along the road and through the gate, and was then very rude to the owner. If my memory serves me aright, we did, however, save his woolshed. The fire had gone past but there was timber stacked beneath it which was burning. We pulled the burnt timber out and extinguished the fire. I think I went back the next day to strain up the fence; but I never apologised for my rudeness.

Lessons C and D: at a fire there is an imperative to get on with putting it out but that doesn't justify rudeness: apologies cost nothing except a little courage and consideration for the other person.

Further west along the road we put out a fire in a knot-hole half-way up a power pole, perhaps the best thing we did that day, otherwise power supplies would have been disrupted by a fallen pole. By the time we arrived at Gill and Ian Buchanan's the fire had already swept through. We saved their overhead tank-stand on wooden poles which had started to burn. Fortunately their house was built so that the walls came down to the ground, preventing fire access beneath it.

In those days when a crop of wheat or barley was harvested it was bagged off the harvester. We came across a farm where the bags had all burnt but there was little we could do to help the owner and his harvesting crew.

Another inconsiderate act was that, in order to get to the fire, I drove through a standing wheat crop, starting another one; however, the crop still had a green bottom and not much damage was done – or perhaps I just did not look back to check.

In one place the fire had bypassed a house but was burning back towards it. Breaking all my own rules about starting from the back of a fire I put the truck in low ratio and crawled over some very large stones as the boys on the back extinguished the fire. The owner wrote me a note of appreciation.

I have deeply imprinted memories of the fire racing towards the Midland Highway and of a Shire truck equipped with tank and pump driving along faster than the men on the back could extinguish the fire so that their efforts were useless. Accompanied by other trucks we went more slowly and put out the fire's left flank but we could not outpace it. The west wind was so strong that the boys were almost blown off the truck.

There were others, including the Shire engineer Ken Middleton, making a valiant attempt to stop the fire crossing the highway.

At one stage I pulled the truck off into a safe place for respite; so that the boys and I could take stock of the situation.

Then we came across a man lying unconscious on the ground. I had no idea what to do so I parked the truck to put him in its shadow. Then someone else came along and whisked him away, crumpled up in the cabin. There may have been a First Aid Centre unknown to me.

Driving home from Geelong after work, June did not know that there had been a fire; because all the ash had blown away! That ends the account of my somewhat inglorious attempts to help at the Lethbridge fire. With various allusions to my activities both on and off the farm it also says a lot about the middle years of my time in charge at *Glenspean*.

9 Fireman Neil

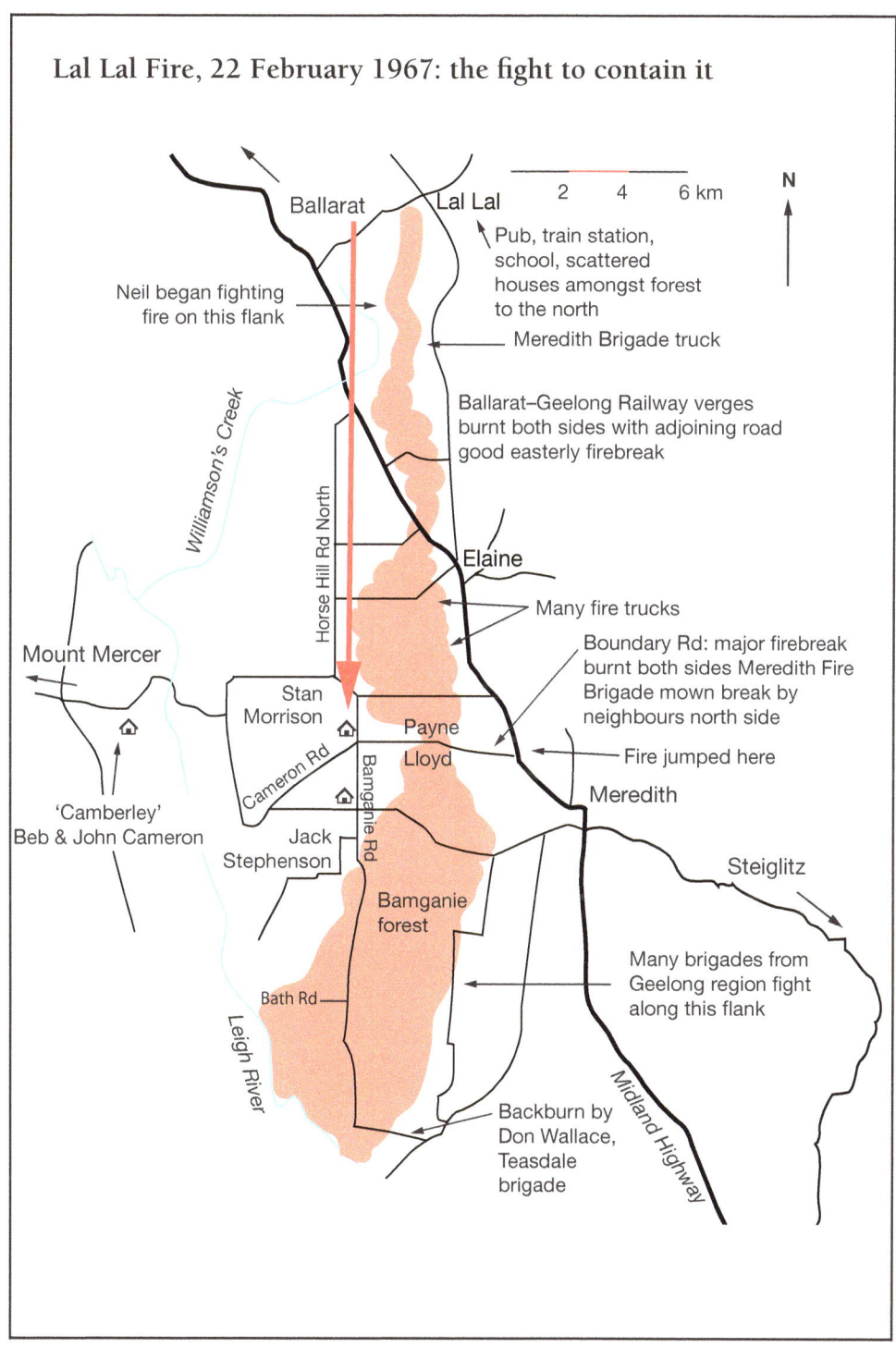

10 After thoughts

Harry Hitchcock from Grenville saved the Woodburn school: cursed by some! Several years later, whilst I was in hospital the chairman of the school committee called a meeting in my absence and voted that the Education Department build a new school. They knew that I had been against the proposal out of deference to Lex McNaughton who had generously built the existing school on a corner of his land. Lex blamed me and refused to be mollified. He ignored the acquisition notice and refused to treat with the department. The land was compulsorily acquired, creating a new allotment; but the department was only offering farmland prices for the 2-acre lot. Consequently, even as Lex's school building was being carted away he was not being paid. I intervened, calling in the local Upper House member Murray Byrne who happened to be the Minister. His department sent a special envoy to deal with Mr McNaughton in person.

We did get a new school. It was lit with electricity when Meredith was connected to the State Electricity Commission in 1978. That was by a group scheme in which I and many others had invested. The interest on the investment paid our electricity bills for years. Although I cannot claim any direct credit for instigating the scheme, I had been lobbying for a power supply for the Meredith district for ages.

On the day of the Lal Lal fire, Ballarat vet Len Fulton had been due to pregnancy test some cows. When a week later he was rescheduled to come and do it, he had to be put off because access was denied due to that section of the Meredith Mercer Road being primed for sealing that day. Many of the older councillors would have been content to spend the Shire's rate money on pick and shovel projects close to home. With others such as Jim Venters I pushed instead for us to use Shire rate revenue to match the generous road grants available at the time – anything up to

10:1. As a result there are many miles of 12-foot wide sealed roads which today still form the basis of the transportation network in the southern section of the Golden Plains Shire, although many of the roads have since been widened or reconstructed.

11 Pioneer Park

One activity with lasting benefit to the people of Meredith was establishment of Pioneer Park on a piece of wasteland adjoining the Recreation Reserve but vested in the Victorian Railways. I moved in the Shire Council that it should seek to have ownership transferred to the Recreation Reserve and to establish a park thereon in celebration of Australia's bicentenary. With slight misgivings the council agreed; the state government did not, but the community proceeded to usurp the land for its own use, for the planting of the park.

We put it to use a dozen years later when we saw in the new millennium. I spent the evening of 31 December 1999 mixing concrete and inviting people to help create a monument to the occasion by placing a stone and with it their hurts of the past and their hopes for the future.

I put the idea of a park to a public meeting where we agreed to create one, government opposition notwithstanding, on the supposition that once established no government would bear the odium of demolishing it or selling it off.

It seems that people are like sheep when it comes to group leadership. On this occasion five of us put our hands up offering to help turn the idea into a reality. The five of us met just once. I prepared a table for the meeting with a copy for each of us. There were columns with the names of the participants across the top, and room for tasks that needed to be undertaken down the side. At the end of the meeting, with the appropriate boxes filled with what everyone was to do, we each had a copy of our own tasks and those of others. There were no further meetings as they were unnecessary. We each had our own things to do without tripping over each other.

My first task was to prepare the ground. I co-opted the Shire grader to level it. We decided on the species of trees that would be planted and drew up a plan of the planting. It had been a dry autumn: Wally and Dutchy watered planting sites marked by stakes on a grid pattern.

The invitation to participate in the planting of the park was given only to residents and their immediate friends or relatives. The bottom pub push were supporters of a contender for my seat on the council, whom I had defeated. They did not attend; yet with a township population of 280 we had 400 people present. With the planting of the park on 1 May 1988 we celebrated Australia's bicentenary; 30 years later our focus might be different.

Proceedings began with the Pony Club re-enacting the arrival of the first explorers. There was a blessing of the project by the three resident clergy and a race for the Meredith Gift. On crutches at the time I was given a generous handicap. When I looked like winning I had to pretend to fall over and poor Mrs McNaughton, fearing for my hip prosthesis, nearly had a fit. We planted our trees, put our names on the stakes and recorded it all. The planting done, we enjoyed lunch of salad and *Glenspean* baby beef butchered by councillor colleague Robert Cooke. We topped up the meal with a large cake provided by Linda. As the late autumn dusk drew on, warmed by a big fire lit by Jim Hynds with logs from the site, we organisers and all our helpers enjoyed the leftovers.

Lyndsay Fink helped me erect the name of the park inscribed by Roy Taylor on a slab of ironbark, one of my strainer posts split lengthwise. (Ironbark is a very durable eucalypt timber: *E. sideroxylon*).

Subsequently the town team suggested asking Meredith families to plant a tree in the adjoining section of Wilson Street, alternately claret and golden ash. The town team paid for the steel, sourced by Don Atherton, from which Kevin Gargan and others made tree guards according to a robust and attractive pattern. These were decorated with motifs appropriate to families who had planted each tree. They were then powder coated green.

Another sheepish activity sponsored by MAA, my short-lived Marvellous Meredith Association, was tree planting in other streets. At my suggestion we planted spotted gums (*Corymbia maculata*) on the lower section of Wilson Street, the idea being to provide wind protection to the southern section of Meredith. A personal activity was to beautify the first block of Read Street running across from the recreation ground, planting various varieties of ornamental cherry plum, which I knew from our garden experience would grow. Others continued the planting past the main street.

Above: watering the bare planting sites, which Wally Koostra and Dutchy Don Van Beusichem volunteered to do; that and watering in the trees when planted

Below: author welcoming the pony club riders pretending to be the first explorers or settlers

Above: author with trees on the truck ready to be planted

Below: author on crutches behind the oldies watching proceedings – Annie McNaughton and Bessie Nolan

Above: the three resident clergy, Rev Don Fairlie, Rev Rob Mitchell and Father Tom Furey leading the half hour of worship, concluding with a skit, Rob acting as one praying and Don acting as God. (Funny but challenging, I recorded)

Below: June immersed in planting a tree whilst Julie Cameron looked on

Above left: before the planting day, author with a pile of colour coded stakes preparing the planting sites according to a plan by Jim Alexander

Above right: Sandy Cameron planting a tree, 1 May 1988

Below: memorial cairn built as we celebrated the millenium, 1999

Top: (Councillor) Robert Cooke and Roy Ritchie carving the steak

Bottom: Food for the multitude: the luncheon table – beef and salad sandwiches with more billy tea and damper; serenaded by Jim Newbury.

Top: planters congregated for an informal ecumenical church service

Above: Andrew Nolan leading in the first ever Bicentennial Gift

Right: Lyndsay Fink boiling the billy and cooking the damper for morning tea/lunch

Above: The winners of the Bicentennial relay: Andrew Cooke, Chris Gargan, John Dixon (Shire secretary) and Kevin Gargan

Below: NC (again!) Cutting the bicentennial cake held by Diane Ritchie and Rosemary Gargan, cooked by Linda Kiernan

Pioneer Park, of lasting value to the Meredith community

All these community activities of mine were 'small beer' compared to the Glenspean Experiment 1953–58, which had launched me on such a positive approach to life. From early days I had been a sheepman on *Glenspean*. As the experiment with cattle proceeded, the farm which I inherited I now began to make my own.

The experiment provided a tangible bridge between my growing-up years and my adult years as a self-reliant farmer, community member and above all as husband. June and I were married towards the close of the experiment. My emancipation from the slavery of self had released to me the ability to speak in public: the experiment gave me a talking point. I came to be in demand to speak on scientific matters bovine.

Epilogue

The Camerons of Glenspean records five generations the Cameron family who have displayed initiative and innovation. It centres on the figure of my father, Neil Wilson Cameron, the story encompassing his grandparents' to his grandchildren's generations. It was he who, with his fiancée Kay by his side, decided to purchase the property *Glenspean* 94 years ago. In marriage they gave birth to my older brother John and to me. In 1948, a quarter of a century into the family's ownership of the property, they took John and me into a partnership, henceforth known as Glenspean Partners.

Glenspean was home to both of us boys until, upon his marriage to 'Beb', John set about establishing a new home on that part of the farm that lies west of the Leigh river. Although John no longer lives at the *Glenspean* homestead he is very much a Cameron of *Glenspean*.

John Cameron, aged 92, leading the 2018 Anzac Day parade

On the day of the fire, as a partner in Glenspean Partners, John was in the *Glenspean* shearing shed classing the wool clip. I had taken time off from my outside duties of mustering the woolly sheep and dipping the shornies in order to help Jeff Cooper unload the kitchen cupboards, which were part of our renovations to the house. June and I were putting our stamp of ownership on *Glenspean*, it being 10 years into our 30 years of living there as a family, before we moved out to make room for the next generation, our son Sandy and his wife Julie.

With the news of fire to the north, there was a general scatter of the shearing-shed personnel. If there was a fire it had to be put out. John Cameron and I both took trucks belonging to Glenspean Partners. Rather than wait for the fire to come to us, I took the initiative to go 20 kilometres north, several kilometres over the Midland Highway, until we met the fire heading south. After several hours, our presence there was no longer effective, and I decided to go back nearer to home.

With John Nolan we fought to save the homestead of Stan Morrison, friend and neighbour. In my father's time he would stop to speak to the Nolans if he met one of them on the road, but otherwise, as Catholics and Protestants, we had little interaction. I had helped to break down that barrier by attending an open-house invitation to John Nolan's birthday party and later on by attending the Requiem Mass for a notable resident. June continued the integrating process by forming an ecumenical choir with John's wife Mary Nolan and with Anglican church member Dawn Macdonald.

Fighting the fire was intensely personal, requiring great concentration. It was personal also because those whose farms were burnt out were members of our own community. Many, such as widow Betty Payne, and George Lloyd, were members of our own vibrant Presbyterian Church, a community within the wider community. I had sat with John Payne but could find no words to help him in his long drawn-out struggle to live.

Half a dozen years later, after a Sunday afternoon of activity with church youth, it being still daylight, I decided to visit Betty Payne and found myself following a police car with two officers in their police caps and immediately sensed the awful truth. I can still see the look of alarm on the faces of Betty and young John as they came out onto the verandah to see what was going on. The police took me to the Ballarat morgue to identify Susan Payne's body.

Community sentiment was that Susan had caused the accident, so I was not entirely popular when I spent the following two days collecting evidence that the man who had collided with her was drunk. One of those policeman had been drinking with him! At the inquest it emerged that the debris on the road indicated that 19-year-old Susie had been on the correct side of the road. A student nurse, she had been returning to Ballarat in order to be ready to scrub for an important operation on the Monday morning.

This story of community involvement leads me to make brief reference to some other fields. A group of Meredith worthies nominated me for a seat on the Bannockburn Shire Council. My first 11 years were not of great note except all councillors greeted each other with a handshake as they entered the meeting and at its end. As a member of the Meredith community I took to greeting everybody by name with a handshake. If I failed to name a person I could see him shrink.

I often became frustrated with the discussion around the table, in which every councillor had his say, but as a group they said nothing. Council was dithering over a possible water supply for the shire. I jumped to my feet and moved that the matter be put in the hands of a committee of three, naming councillors Boardman, Cooke and Harvey plus the shire engineer, it being given the power to act! And so we, with parts of the Leigh Shire, achieved a water supply for the six townships and for farmers adjoining the pipelines.

Council used to meet monthly at 10.00 am, take lunch at noon and continued to late afternoon, when the drinkers amongst the councillors would stay to partake of the shire president's liquid refreshments and the non-drinkers would go home. With June's help in my first presidency I provided a light meal to go with the drinks. Helped by the likes of Jim Venters, we introduced a more workmanlike protocol to council meetings so that we elected to not meet until afternoon. This made it possible for people who had other employment to sit on the council and attend meetings. June's light snack devolved into a council dinner during which we were all able to enjoy each other's company.

During my second presidency we asked the State Governor, Sir Henry Winnekie, to open the new shire offices. We could have had a bald half-hour ceremony with boring speeches in the presence of a few community representatives. I insisted that we issue an open invitation to the proceedings, in response to which half the shire population attended. For entertainment I asked the children from the several schools within the shire to provide a pageant depicting our life as a community. Formal opening speeches followed. Then I provided afternoon tea for all.

Epilogue

Opening the new Shire of Bannockburn offices; children presenting a pageant for the governor Sir Henry Winneke

After 33 years, with me in the chair as 'mayor' for my penultimate year before retiring, Premier Jeff Kennett gave us all the sack to make way for municipal amalgamations, and that was the end of voluntary local governance. It was also the end of our direct community involvement as Camerons of *Glenspean*, for June and I were moving out of *Glenspean* to make way for generation five. My swansong was the creation of Pioneer Park as a celebration of Australia's bicentennary in 1988, and in the park so created 11 years later we farewelled the old millennium and welcomed the new. Fittingly, Pioneer Park is the final chapter in our book.

Chapter 9 Fireman Neil contains many allusions to my community activity. The Lal Lal fire described in it is a long-forgotten memory except for those of us still living who were intimately involved in the action; but the establishment of Pioneer Park is of lasting value to the Meredith community. This is a brief sketch of my time as a fourth-generation Cameron of *Glenspean*.

So we end where we began the book, proclaiming Sandy and Julie Cameron as fifth-generation Camerons; they are most definitely showing initiative and innovation in their conduct of the Meredith Dairy.

Published by GLENSPEAN: mg.house@bigpond.com; 0429 415 973
Profits from the sale of this book will go to Oxfam, The Fred Hollows Foundation and CBM
Text and photographs copyright © Neil G. Cameron 2018; illustrations (pp. 61–2) by the author
Editing: Bet Moore
Design, layout, diagrams and maps: Nan McNab

www.ingramcontent.com/pod-product-compliance
Lightning Source LLC
Chambersburg PA
CBHW040325300426
44112CB00021B/2877